FOOD PROTEINS
AND LIPIDS

ADVANCES IN EXPERIMENTAL MEDICINE AND BIOLOGY

Recent Volumes in this Series

FOOD PROTEINS AND LIPIDS

Edited by

Srinivasan Damodaran
University of Wisconsin – Madison
Madison, Wisconsin

Springer Science+Business Media, LLC

Library of Congress Cataloging-in-Publication Data

Food proteins and lipids / edited by Srinivasan Damodaran.
       p.   cm. -- (Advances in experimental medicine and biology ; v.
415)
    "Proceedings of the John E. Kinsella Memorial Symposium on Food
Proteins and Lipids, held August 22-23, 1995, in Chicago, Illinois"-
-T.p. verso.
    Includes bibliographical references and index.

    1. Proteins in human nutrition--Congresses. 2. Lipids in human
nutrition--Congresses. 3. Food--Protein content--Congresses.
4. Food--Lipid content--Congresses. 5. Kinsella, John E., 1938--
-Congresses. I. Damodaran, Srinivasan. II. John E. Kinsella
Memorial Symposium on Food Proteins and Lipids (1995 : Chicago,
Ill.) III. Series.
QP551.F779 1997
612.3'98--dc21   97-6056
                                                            CIP

Proceedings of the John E. Kinsella Memorial Symposium on Food Proteins and Lipids, held August 22 – 23, 1995, in Chicago, Illinois

DOI 10.1007/978-1-4899-1792-8

© 1997 Springer Science+Business Media New York
Originally published by Plenum Press, New York in 1997
MyCopy version of the original edition 1997

10 9 8 7 6 5 4 3 2 1

PREFACE

John E. Kinsella, Dean of the College of Agricultural and Environmental Sciences at the University of California–Davis, passed away on May 2, 1993, at the age of 55. In August 1995, former students and post-doctoral fellows of Dr. Kinsella met at the American Chemical Society National Meeting in Chicago to convene a Symposium on Food Proteins and Lipids to honor Dr. Kinsella's enormous contribution to the field of food science and nutrition. This book is a collection of papers presented at that symposium.

A native of Ireland, Dr. Kinsella received his bachelor's degree in agricultural sciences in 1961 from the University of Dublin. He received his master's degree in biology in 1965 and a doctorate in food chemistry in 1967 from Pennsylvania State University. He joined the Food Science faculty at Cornell University in 1967. While at Cornell, he served as Chair of the Department of Food Science from 1977–1985 and Director of the Institute of Food Science from 1980–1987. He was designated Liberty Hyde Bailey Professor of Food Biochemistry in 1981, a Fulbright Fellow in 1983, and was selected as the General Foods Distinguished Professor of Food Science in 1984. He was named a Leading Professor in the State University of New York, the highest professorial honor in the SUNY system. In 1990 he joined the University of California at Davis as Dean of the College of Agricultural and Environmental Sciences.

Dr. Kinsella won many honors and awards, including the Babcock-Hart Award in 1987 from the Institute of Food Technologists, the Atwater International Award in 1988 from USDA, the Distinguished Lectureship Award from the Philadelphia section of the Institute of Food Technologists and the Outstanding Professor Award from IFT in 1989, the Spencer Award from the American Chemical Society in 1990, and the Stephen S. Chang Award for distinguished research in lipid biochemistry in 1991 from the American Oil Chemists' Society.

Dr. Kinsella published over 500 research papers, chapters, and reviews, and held several patents. He trained more than 65 M.S. and Ph.D. students and about 46 post-doctoral fellows and research associates. All contributing authors of this book are former colleagues of Dr. Kinsella.

The encouragement and support provided by the Division of Agricultural and Food Chemistry of the American Chemical Society in organizing the symposium is gratefully acknowledged. Financial support from General Mills, Solvay Enzymes, Universal Foods Corporation, Campbell Soup Company, Kraft Food Ingredients, Central Soya Company, The Pillsbury Company, and the Academic Press is also gratefully acknowledged.

Srinivasan Damodaran

CONTENTS

MOLECULAR DESIGN OF SOYBEAN GLYCININS WITH ENHANCED FOOD QUALITIES AND DEVELOPMENT OF CROPS PRODUCING SUCH GLYCININS

S. Utsumi,[1]* T. Katsube,[1]** T. Ishige,[2] and F.Takaiwa[2]

[1]Research Institute for Food Science
Kyoto University
Uji, Kyoto 611, Japan

[2]National Institute of Agrobiological Resources
Tsukuba, Ibaraki 305, Japan

INTRODUCTION

Soybean (*Glycine max* L.) protein has a function to lower cholesterol level in human serum (Kito et al., 1993) and is one of the best plant food proteins in terms of nutritional and organoleptic qualities. However, it is usually inferior in these respects to animal proteins. For example, the amino acid composition of soybean protein does not satisfy infant requirement: sulfur containing amino acids are deficient. Improvement of nutritional value and functional properties of soybean proteins is one of major objectives in the food industry (Kinsella, 1979). Soybean proteins consist of two major components, glycinin and β-conglycinin (Derbyshire et al., 1976). Of these two proteins, glycinin is superior to β-conglycinin with regard to nutritional value (Millerd, 1975) as well as functional properties (Kinsella, 1979). Five subunits are identified as constituent subunits of glycinin and classified into two groups: group I (A1aB1b, A1bB2, A2B1a) and group II (A3B4, A5A4B3) (Utsumi, 1992). The subunits belonging to group I have better nutritional value than those of group II. Therefore, a group I subunit of glycinin is a suitable target for such improvement to create an ideal food protein. Protein engineering appears to be a promising method in achieving improvement of glycinin qualities because the primary sequence of glycinin can be modified consciously and systematically.

* Corresponding author.
**Present address: Shimane Women's College, Matsue, Shimane 690, Japan.

Glycinin is composed of six subunits, each of which consists of an acidic and a basic polypeptide that are linked by a disulfide bridge (Badley et al., 1975; Mori et al., 1979; Staswick et al., 1981, 1984). The constituent subunits of glycinin are synthesized as a single polypeptide precursor (preproglycinin) consisting of covalently-linked acidic and basic polypeptides together with a signal sequence (Staswick et al., 1981; Barton et al., 1982; Tumer et al., 1982). The signal sequence is removed cotranslationally in the endoplasmic reticulum (ER), and the resultant proglycinin subunits assemble into trimers of about 8S (Barton et al., 1982; Tumer et al., 1982; Chrispeels et al., 1982). These complexes are targeted from the ER to the vacuoles, where a specific posttranslational cleavage occurs (Nielsen, 1984), where they form protein bodies. The cleavage results in mature subunits, each of which consists of an acidic and a basic polypeptide and they assemble into hexamers of about 12S (Barton et al., 1982; Tumer et al., 1982; Chrispeels et al., 1982). Finally glycinins accumulate in a highly packed state in protein bodies. Molecular assembly, targeting from the ER to the vacuoles and accumulation in protein bodies depend on topogenic information contained in the glycinin molecule. Therefore, it is essential to find what kinds of modification by protein engineering glycinin molecules can tolerate without misfolding when creating novel soybean plants that can produce modified glycinins having improved food function (Utsumi, 1992).

It is necessary to evaluate whether protein-engineered glycinins are able to form a conformation similar to that of native glycinin, to exhibit expected functional properties and to accumulate in the protein bodies of plant cells before the modified genes are transferred to the soybean plant and other crops. To evaluate these points, establishment of a high-level expression system of glycinin cDNA in *Escherichia coli* and a transformation system using tobacco is desired.

EXPRESSION OF SOYBEAN GLYCININ CDNA IN ESCHERICHIA COLI

Expression of glycinin cDNA in *E. coli* was attempted using expression vector pKK233-2. It was difficult to detect the translational products from the cDNAs encoding A1aB1b or A2B1a preproglycinin in *E. coli* cells or the media (Utsumi et al., 1987). This phenomenon was supposed to be due to the presence of the hydrophobic signal sequence which was not cleaved in *E. coli* and disturbed the folding of the expressed protein to the correct conformation. This was confirmed by the stepwise deletions of DNA sequence encoding the signal sequence and the mature NH_2-terminal region from A1aB1b cDNA. Although no expressed proteins from the cells harboring an expression plasmid for a proglycinin homologue protein containing the five amino acids of the signal sequence were detected in either the cells or the medium, the products accumulated as soluble proteins in the cells harboring expression plasmids for proglycinin homologue proteins having less than five amino acids of the signal sequence and lacking one to eleven amino acids from the mature NH_2-terminal (Utsumi et al., 1988a). Therefore, it is strongly suggested that the folding of the expressed proteins, with more than five amino acids of the signal sequence, may be disturbed and this could make the proteins susceptible to proteinase digestion. Thus, the deletion of the signal sequence coding region from the glycinin cDNA is essential for the expression in *E. coli*. The highest expression level was observed with the expression plasmid pKGA1aB1b-3. The expressed protein A1aB1b-3 from this plasmid lacks the NH_2-terminal three amino acids and has the initiation methionine (Utsumi et al., 1988a).

To attain high-level expression, we attempted to increase the copy number of the expression plasmid using a runaway vector (Uhlin et al., 1979) or the replication origin of pUC vector (Miki et al., 1987) and to change *E. coli* strain and the distance between the Shine-Dalgarno sequence and the initiation codon. However, these attempts did not increase the level of A1aB1b-3 expression (Kim et al., 1990a). Then the culture conditions of *E. coli* strain JM105 harboring pKGA1aB1b-3 were changed. By controlling the culture conditions to 37 and 90 strokes/min, the soluble expressed protein was obtained at a high-level corresponding to 20% of the total bacterial proteins (Kim et al., 1990a). Even at such a high level, the expressed protein did not form inclusion bodies, suggesting that A1aB1b-3 proteins have a tendency to form proper conformation in *E. coli*.

E. coli cells do not have enzymes responsible for cleavage of proglycinin to a mature form (Utsumi et al., 1988a). Therefore, the expressed proteins from pKGA1aB1b-3 accumulate as proglycinins in *E. coli* . The proglycinin A1aB1b-3 were verified to assemble into trimers as observed in the ER of soybean cells by a sucrose density gradient centrifugation, and to have the secondary structure similar to that of glycinin by circular dichroism (Kim et al., 1990a). It is known that some misfolded proteins have a tendency to form insoluble inclusion bodies in *E. coli* (Williams et al., 1982; Shoemaker et al., 1985). These facts together with the observation that the folding of proglycinin homologue proteins with more than five amino acids of the signal sequence were disturbed and rendered to be susceptible to proteinase digestion propose the following three criteria for judging formation of proper conformation of protein-engineered proglycinin: (i) accumulation as a soluble protein should be observed and solubility should be comparable with that of globulin, (ii) there must be self-assembly into trimers, and (iii) high-level expression in *E. coli*. Recently, we demonstrated that stability of conformation is an alternative to the third criterion (Gidamis et al., 1995). In addition, A1aB1b-3 exhibits the fundamental properties such as cryoprecipitation and calcium-induced precipitation and the functional properties such as heat-induced gelation and emulsification as glycinin does (Kim et al., 1990a). Therefore, the *E. coli* expression system established here can be used for the evaluation of the formation of proper conformation and the food functions of protein-engineered glycinin.

The cDNA sequence encoding A1aB1b preproglycinin containing 3'- and 5'-noncoding regions was placed under control of the repressible acid phosphatase promoter *PHO5* of the yeast *Saccharomyces cerevisiae* in an expression vector pAM82. The signal sequence of the expressed protein was correctly recognized and processed at the same site as in soybean by the yeast processing system (Utsumi et al., 1988b; 1991). The expressed proteins accumulated as proglycinin in the cells (the expression level was ~5 % of the total yeast proteins or 30-40 mg per liter of culture), and most (~90 %) of the expressed proteins were insoluble due to their interaction with intracellular components at the acidic polypeptide region (Utsumi et al., 1991). Therefore, the *E. coli* expression system is superior to the yeast system for protein engineering of glycinin.

DESIGN OF MODIFIED GLYCININS WITH ENHANCED FOOD FUNCTIONS

When attempts are made to improve the nutritional and functional properties of glycinin by protein engineering, the following two problems should be considered: (i) which regions of glycinin are susceptible to modifications by protein engineering? and (ii) what kinds of modifications are employed? Wright (1988) aligned the amino acid sequences to maximize the homology among the 11S globulins from various legumes and nonlegumes, and suggested that they comprised a series of alternating conserved and variable regions. The existence of five variable regions, I-V, was suggested (Figure 1A) (Wright, 1988). All variable regions exist in the hydrophilic regions, which suggests that they are located on the surface of the protein. The variable regions probably have little function in forming and maintaining the glycinin structure and may tolerate modification. The relationship between the structure and the functional properties of glycinin may answer the second question relating to the kind of modification that should be employed. Nakamura et al. (1984)

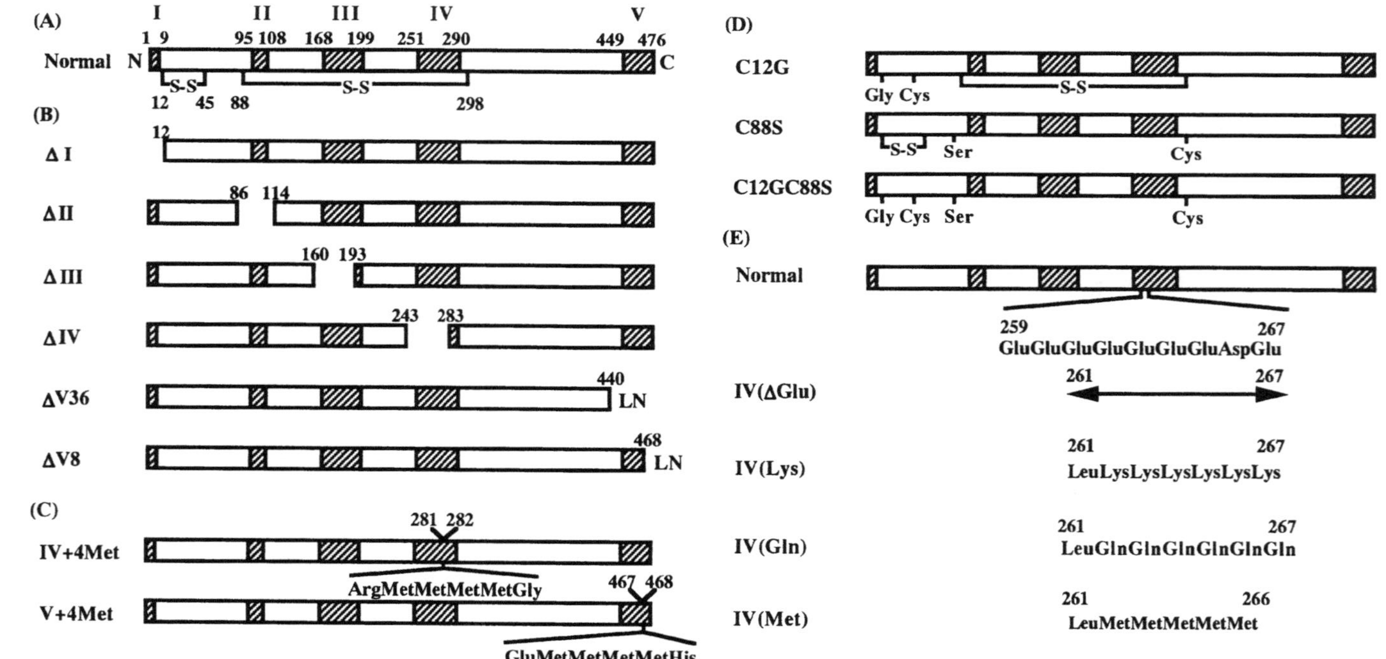

Figure 1. Schematic representation of the normal and modified proglycinins. Open and hatched areas are conserved and variable regions, respectively. N and C represent NH_2- and COOH-termini, respectively. The numbers of residues from the NH_2-terminus for the variable regions I-V are shown above the alignment. (A) The variable and conserved regions of the normal proglycinin A1aB1b. The positions of the disulfide bonds are indicated. (B) Deletion mutants lacking each variable region. (C) Insertion mutants having tetramethionines. (D) Disulfide bond-deleted mutants. (E) Deletion and substitution mutants having higher isoelectric points. [Adapted from Kim et al. (1990b), Utsumi and Kito (1991), Utsumi (1992), and Katsube et al. (1994), with permission of the authors and publishers.]

4

studied the relationships between the structure at the subunit level and heat-induced gelation of glycinin and proposed that the heat instability of the constituent subunits was related to the heat-induced gel-forming ability. On the other hand, attachment of fatty acid to glycinin increased its emulsifying properties (Utsumi and Kito, 1991). This indicates that hydrophobicity is an important factor in the emulsifying properties of glycinin. Kato and Yutani (1988) reported that the surface properties of a protein depend on the conformational stability: the more unstable, the higher the emulsifying properties. Disulfide exchange plays an important role in the formation of heat-induced gel (Mori et al., 1982), and the number and the topology of free sulfhydryl residues are closely related to the heat-induced gel-forming ability and the gel properties of glycinin (Nakamura et al., 1984; Utsumi, 1992). These facts suggest that (i) partial or complete removal of the variable regions, (ii) insertion of plural hydrophobic amino acids into the variable regions, and (iii) the substitution of a sulfhydryl residue involved in disulfide bond formation for another amino acid may be powerful methods of improving the food qualities of glycinin. Such modifications may induce a strengthening of the hydrophobicity, change of the number and the topology of free sulfhydryl residues, and the destabilization of the glycinin molecule. On the other hand, improvement of the nutritional values of glycinin can be achieved by fortification of the limiting essential amino acid. We designed fifteen modified glycinins shown in Figure 1. Rationale of modifications is classified into four groups.

Deletion of Each Variable Region

Each variable region has strong hydrophilic nature. Removal of each variable region results in strengthening relative hydrophobicity and partial destabilization of the glycinin molecule. Consequently, improvement of the heat-induced gel-forming and emulsifying abilities could be expected. Therefore, deletion mutants lacking each variable region were designed as shown in Figure 1B. ΔI lacks NH_2-terminal eleven amino acids, ΔII from the 87th to the 113th, ΔIII from the 161st to the 192nd, ΔIV from the 244th to the 282nd, ΔV36 from the 441st to the COOH-terminus, and ΔV8 from the 469th to the COOH-terminus (Kim et al., 1990b). The last two mutant glycinins have two extra amino acids (Leu-Asn) at their COOH-terminus derived from the universal terminator sequence.

Insertion of Tetramethionines into Variable Regions

The limiting essential amino acid of soybean proteins is methionine, which has hydrophobic nature. Therefore, insertion of contiguous plural methionines into variable regions results in improvement of both the nutritional and the functional qualities. Saalbach et al. (1990) introduced multiple methionine codons into field bean legumin gene by frame shift and site directed mutagenesis at DNA region encoding COOH-terminal region of legumin, but modified legumin was not accumulated in the transgenic plant seeds. This suggests that modifications causing a fairly big change of structural characteristics are not suitable for improvement of food functions. Then, insertion mutants of IV+4Met and V+4Met were designed (Figure 1C) (Kim et al., 1990b). IV+4Met has Arg-Met-Met-Met-Met-Gly between Pro281 and Arg282. V+4Met has Glu-Met-Met-Met-Met-His between Pro467 and Gln468. The insertion results in change of hydropathy profile; hydrophobicity of the insertion site increases (Figure 2).

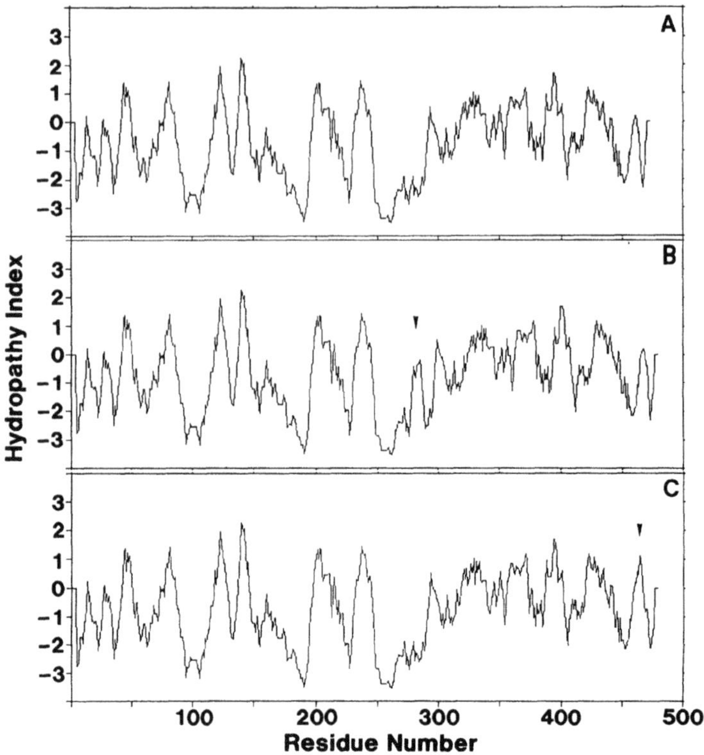

Figure 2. Hydropathy profiles of the normal and modified proglycinins: (A) normal proglycinin; (B) IV+4Met; (C) V+4Met. The arrow heads indicate the insertion site.

Deletion of Disulfide Bond(s)

Each of glycinin constituent subunits has two or three disulfide bonds (Utsumi et al., 1996). Two disulfide bonds, one at residues 12-45, which is in the acidic polypeptide region, and the other at 88-298, which is between the acidic and the basic polypeptide regions, are identified in A1aB1b subunit (Staswick et al., 1984; Utsumi, 1992). Disulfide bond-deleted mutants C12G and C88S were designed, where Cys12 and Cys88 were substituted with Gly and Ser, respectively (Figure 1D) (Utsumi et al., 1993a). As a result of this modification, C12G and C88S have a new free cysteine residue at 45 and 298, respectively. We also designed C12GC88S having two new free cysteine residues at 45 and 298, where both Cys12 and Cys88 were substituted.

Deletion and Substitution of Polyglutamic Acid Sequence

The variable region IV is the most variable among the five variable regions (Wright, 1988) and termed the hypervariable region (Argos et al., 1985). This region is rich in hydrophilic and negatively charged amino acid residues. The sequence between the positions of 259 and 267 is composed of one aspartic acid and eight glutamic acid residues. We termed this sequence the polyglutamic acid sequence. Soybean proteins do not fully exhibit their functional properties in the acidic range because their solubility decreases due to the effect of their isoelectric points, which also lie within the acidic range (Kinsella, 1979). Soybean proteins are therefore rarely used for acidic foods such as mayonnaise and yogurt. Therefore, the alteration of the net electric charge of the polyglutamic acid sequence would result in improvement of functional properties of glycinin in the acidic range. Then, deletion mutant of IV(ΔGlu) and substitution mutants of IV(Lys), IV(Gln), and IV(Met) were designed (Figure 1E) (Katsube et al., 1994). IV(ΔGlu) lacks glutamic and aspartic acid residues from the 261st to the 267th. The glutamic and aspartic acid residues from the 261st to the 266th or the 267th are substituted with LeuLysLysLysLysLysLys, LeuGlnGlnGln-GlnGlnGln, and LeuMetMetMetMetMet in IV(Lys), IV(Gln), and IV(Met), respectively. Substitution of this sequence with polyglutamine and polylysine sequences brings about no change of the hydropathy profile, but the deletion and the substitution with polymethionine sequence significant change similarly to those of IV+4Met and V+4Met shown in Figure 2 (Katsube, et al., 1994).

EVALUATION OF PROPERTIES OF MODIFIED PROGLYCININS

To achieve development of crops producing modified glycinins with enhanced food qualities, the modified glycinins should be able to assume the correct conformation. In order to evaluate this point, *E. coli* expression plasmids for the modified glycinins designed here were constructed using pKGA1aB1b-3 for the normal proglycinin A1aB1b-3 except for ΔI. The expressed proteins from these expression plasmids accumulate as modified proglycinins in analogy with the normal one. The nucleotide sequences in the vicinity of the translation initiation site and the promoter of each expression plasmid are the same as those of pKGA1aB1b-3. Therefore, the efficiencies of transcription and translation are expected to be identical to each other.

Three criteria described above for judging formation of proper conformation were applied to all the modified proglycinins expressed in *E. coli* cells harboring individual expression plasmids. Among the modified proglycinins, ΔI, ΔV8, IV+4Met, V+4Met, C12G, C88S, C12GC88S, IV(ΔGlu), IV(Lys), IV(Gln), and IV(Met) satisfied the three criteria (Kim et al., 1990b; Utsumi et al., 1993a; Katsube et al., 1994). Therefore, we concluded that these eleven modified proglycinins can form a conformation similar to that of the native proglycinin.

Purification of the eleven modified proglycinins were carried out by ammonium sulfate fractionation and Q-Sepharose column chromatography. All the modified proglycinins except C12GC88S were purified to near homogeneity. C12GC88S was easily degraded during dialysis against column buffer composed of 35 mM potassium phosphate (pH 7.6), 0.15 M NaCl, 10 mM 2-mercaptoethanol, 1.5 mM PMSF, 1 mM EDTA, and 0.02 % NaN3 although it was stable in the same buffer except containing 0.4 M NaCl, indicating that the conformational integrity of C12GC88S is susceptible to attack of proteinase at low ionic strength.

Normal proglycinins could be crystallized from 0.1 M Tris-HCl buffer (pH 7.6) by the dialysis equilibrium method (Utsumi et al., 1993b). Crystallization of the modified proglycinins was attempted by the same method. Although C88S and IV(Lys) did not form crystals by this method, others formed crystals at similar conditions (Gidamis et al., 1994; Katsube et al., 1994). C88S and IV(Lys) also formed crystals by hanging drop vapor diffusion method with PEG6000 as the precipitant. Thus, all the modified proglycinins which were able to be purified could form crystals, confirming the conclusion that these modified proglycinins can form a proper conformation similar to that of the native proglycinin.

Isoelectric points of the modified proglycinins IV(ΔGlu), IV(Lys), IV(Gln), and IV(Met) were measured by isoelectric focusing as follows, respectively: 6.6, 7.2, 6.5, and 6.4 (Katsube et al., 1994). These values are significantly higher than that (5.6) of the normal proglycinin. Especially, the isoelectric point of IV(Lys) changed by 1.6 unit from 5.6 to 7.2. Thus, these modified proglycinins are expected to exhibit good functional properties in the acidic range.

The modified proglycinins ΔI, ΔV8, IV+4Met, V+4Met, C12G, and C88S were purified in large quantities, and their functional properties (gelation and emulsification) were compared with those of the native glycinin and the normal proglycinin (Kim et al., 1990b; Utsumi et al, 1993a). All the modified proglycinins examined here formed gels by boiling for 30 min. The normal proglycinin formed gels having similar hardness to that of the native glycinin. The gels from ΔI, IV+4Met, V+4Met, and C88S had higher hardness than the native glycinin gels, although C12G gels had similar hardness at higher protein concentration (but did not form gels at lower protein concentration) and ΔV8 gels had lower hardness. A noteworthy finding is that C88S could form hard gels at a low protein concentration where the native glycinin formed very soft gels (Utsumi et al., 1993a). On the other hand, all the modified proglycinins examined here exhibited higher emulsifying activities than the native glycinin. Especially, ΔV8 and V+4Met exhibited twice the value as compared to the native glycinin.

The evaluation of the food functions of the modified proglycinins examined here were summarized in Table I. Most of the modified proglycinins exhibited better properties in at least one function than the native glycinin and the normal proglycinin. Especially, IV+4Met and V+4Met exhibited better properties in heat-induced gel-forming and emulsifying abilities as well as nutritional value. Thus, it is possible to say that IV+4Met and V+4Met are ideal food proteins.

Table 1. Summary of evaluation of food functions of modified proglycinins

Modified proglycinin	Emulsification[1]	Gelation[1]	Isoelectric point[2]	Nutritional value[2]
ΔI	Better	Better	N.D.[3]	–[4]
ΔV8	Much Better	Poorer	N.D.	–
IV+4Met	Better	Better	N.D.	Better
V+4Met	Much Better	Better	N.D.	Better
C12G	Better	Changed	N.D.	–
C88S	Better	Much Better	N.D.	–
IV(ΔGlu)	N.D.	N.D.	Higher	–
IV(Lys)	N.D.	N.D.	Much Higher	–
IV(Gln)	N.D.	N.D.	Higher	–
IV(Met)	N.D.	N.D.	Higher	Better

[1]Comparison with the native glycinin. [2]Comparison with the normal proglycinin. [3]Not done. [4]No change.

EVALUATION OF TARGETING, PROCESSING, AND ASSEMBLY OF MODIFIED GLYCININS USING TOBACCO PLANTS

It is possible to evaluate by using *E. coli* expression system whether the modified proglycinins can form a proper conformation similar to that of the normal proglycinin and exhibit expected-functional properties. However, evaluation of targeting into vacuoles (protein bodies), processing into pro- and mature-forms, assembly into hexamers, and accumulation in seeds is impossible by *E. coli* system. For this, examination using transgenic plant is required. Tobacco is one of the easiest plant to get transgenic plant. We attempted to investigate the targeting, processing, assembly and accumulation of the modified glycinins IV+4Met and V+4Met with better functions than the native glycinin and the other modified one by using tobacco.

Expression of Modified Glycinins by Cauliflower Mosaic Virus 35S Promoter

The cDNAs for the normal glycinin, IV+4Met, and V+4Met were placed under the control of the cauliflower mosaic virus (CaMV) 35S promoter in the binary vector pBI121, and then introduced into the genome of tobacco by *Agrobacterium*-mediated transformation (Utsumi et al., 1993c). Five independent plants for each construct were regenerated. Proteins extracted from seeds, leaves, and stems of regenerated plants were assayed immunologically for the presence of glycinin. Although expression levels varied among independent regenerated plants, the normal and modified glycinins were detected in each tissue. Generally, the highest expression levels were similar among the normal and modified glycinins in any tissues, thus around 0.1 % of total proteins of each tissue (Table II). The proteins extracted from seeds (plants 8, 106, 207), leaves (plants 10, 104, 207), and stems (plants 6, 104, 207) were analyzed for processing to pro- and mature-form and assembly by sucrose density gradient centrifugation and SDS-PAGE. The results indicate that processing

Table 2. Expression levels of normal and modified glycinins in seeds, leaves and stems of individual transgenic tobacco plants.[a]

	Normal					IV+4Met					V+4Met				
	6	8	10	11	15	104	106	107	108	113	201	203	207	209	220
Seeds	++	+++	++	−	+++	++	+++	++	+	+++	+++	+++	+++	++	−
Leaves	+++	−	+++	+++	−	+++	−	+++	−	−	+++	−	+++	−	++
Stems	++	−	−	++	−	+	+	−	−	−	+++	−	++	−	−

+++, > 0.05 %; ++, > 0.01 %; and +, < 0.01 % of total soluble proteins; −, not detectable.
[a] From Utsumi et al. (1993c), with permission of the authors and publisher.

of any preproglycinins to pro- and mature-form occurred in any tissue and any mature glycinins assembled into hexamers in the seeds, although assembly in the leaves and stems were obscure. It is known that the processing enzyme responsible for post translational cleavage of pro-form of glycinin-type proteins exists in the protein bodies derived from vacuoles (Hara-Nishimura et al., 1991; Scott et al., 1992). Therefore, the fact that the normal and modified glycinins were processed to mature-form indicates that they targeted to the protein bodies in seeds or to the vacuoles of leaves and stems. Consequently, we can say that the modifications introduced into IV+4Met and V+4Met do not disturb targeting, processing and assembly, indicating that it is possible to create novel crops (soybean, rice, potato, spinach, etc.) producing modified glycinins (Utsumi et al., 1993c).

Expression of Modified Glycinins by Rice Glutelin Promoter

The major storage protein of rice is glutelin which accounts for around 80 % of the rice storage proteins. Glutelin is a macromolecule composed of disulfide-bonded subunits; the fundamental structures of the constituent subunits are similar to those of glycinin subunits (Utsumi, 1992). Rice proteins are deficient in lysine and rich in sulfur-containing amino acids, and soy proteins are vice versa. Therefore, we can expect harmonic accumulation of glycinin with glutelin, resulting in improvement of nutritional value and endowment of functional properties such as gel-formation and emulsifying abilities. The expression of glutelin genes in rice is restricted to the endosperm tissue whereas glycinin genes are specifically expressed in the embryonic tissue of soybean seeds. As a part of a program to develop a transgenic rice producing modified glycinins, we attempted to know whether glycinin can accumulate at a high-level specifically in the endosperm tissue of transgenic tobacco seed under the control of the promoter of glutelin gene, since tobacco seeds substantially retain the endosperm tissue.

The normal and modified glycinin cDNAs were fused to the 5' flanking region (1320-bp) of the glutelin *GluB-1* gene (Takaiwa et al., 1991), and then inserted into the binary vector pBI101. The chimerical constructs were introduced into tobacco genome by *Agrobacterium*-mediated transformation. Several independent plants for each construct were regenerated (Takaiwa et al., 1995). Proteins extracted from their seeds were assayed immunologically for the presence of glycinin. Both the normal and modified glycinins accumulated and the levels were, on average, more than 2 % of total seed proteins. This expression level is comparably higher than those reported for other systems: 0.003 - 2 % of total seed proteins (Beachy et al., 1985; Sengupta-Gopalan et al., 1985; Higgins et al., 1988; Williamson et al., 1988; Robert et al., 1989; Bogue et al., 1990). In order to know whether such a high-level expression is characteristics of the combination of glutelin promoter and glycinin gene, twenty three and thirty four independent plants were regenerated for the normal glycinin and V+4Met, respectively, and then the expression levels were analyzed immunologically. In both cases, more than 60 % of plants accumulated the normal and modified glycinins at the level of > 1 % of total seed proteins in dry seeds (Figure 3). The highest level was about 4 % for both the normal and V+4Met. We also observed the same highest expression level for IV+4Met. This highest level was more than 40-fold higher when compared with that directed by CaMV 35S promoter. Therefore, glutelin promoter is suitable for the expression of glycinin. On the other hand, there was low plant-to-plant variability in the accumulation levels among independent transformants. There is a possibility that scaffold attachment region to reduce the positional effect (von der Geest et al., 1994) is included in the 1.3 kb of the 5' flanking region of *GluB-1* gene. Synthesized normal and modified proteins were processed into mature-forms, and assembled into hexamers. This suggests that they were targeted into the protein bodies. In fact, immunogold electron microscopy of the seed cells of transgenic tobacco indicated that accumulation of the normal and modified glycinins was confined to the matrix of the protein bodies. However, about half of the synthesized normal and modified glycinins were limited-proteolyzed and assembly into hexamers was insufficient. Similar degradation products were

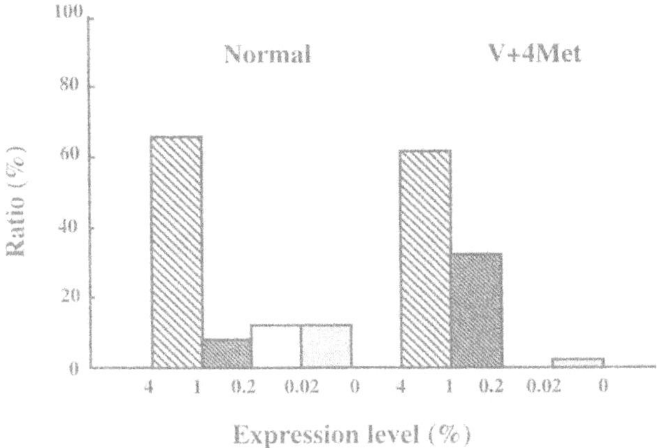

Figure 3. Frequency of expression levels of the normal and modified proglycinins in independent transgenic tobacco seeds. Expression level is % of total seed proteins.

observed often in heterologous transgenic plants, when storage protein genes were expressed, regardless of the promoter. Immunological tissue print of the seed sections of transgenic tobacco revealed that the synthesized normal and modified glycinins were confined to the endosperm tissue surrounding the embryo (Figure 4) (Takaiwa et al., 1995). This accumulation pattern is consistent with the GUS staining pattern directed by the *GluB-1* promoter (unpublished results). Taken together, it can be concluded that the glycinin gene is specifically expressed in maturing seed and its translated products are highly accumulated in the endosperm tissue under the control of rice glutelin promoter. It was further demonstrated that accumulation pattern was not altered by the modification introduced into IV+4Met and V+4Met. These observations strongly suggest that development of rice producing modified glycinins with enhanced food functions is possible (Takaiwa et al., 1995), and we are now trying it and getting good results.

DEVELOPMENT OF CROPS PRODUCING MODIFIED GLYCININS

An easy method to get transgenic plants for some important crops has been established. Potato is one of the easiest important crops among them. The major storage protein of potato is patatin which contributes up to 40 % of the water soluble protein fraction (Racusen and Foote, 1980). This protein is encoded by class I genes, a multigene family, which are specifically expressed in tubers (Pikaard et al., 1987).

The promoter sequence (983-bp) of a class I patatin gene was isolated by PCR method. The normal and modified glycinin cDNAs were joined with the patatin promoter and inserted into the binary vector pBI101. The chimeric constructs were introduced into the genome of potato by *Agrobacterium*-mediated transformation. Six independent plants for each construct were regenerated (Utsumi et al., 1994). Proteins extracted from their tubers were assayed immunologically for the presence, processing, and assembly of the expressed glycinins in a similar manner as employed for tobacco seeds. The highest expression levels were similar

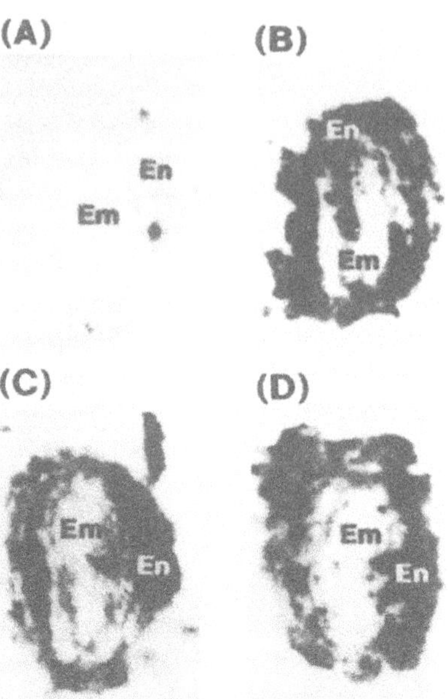

Figure 4. Histochemical localization of the expressed glycinins in mature seeds. Prints of the seed section of untransformed (A), normal (B), IV+4Met (C), and V+4Met (D) were treated with anti-glycinin serum, followed by a goat anti-rabbit IgG-alkaline phosphatase conjugate. Em, embryo; En, Endosperm. [From Takaiwa et al., (1995), with permission of the authors and publisher.]

among the three gene constructs and were 0.5 to 1.0 % of total tuber soluble proteins of tubers. The spatial expression of the chimeric genes was examined by in situ immunocytochemistry. The expression of the normal and modified glycinins was detected in the parenchyma cells but not in the periderm cells. These results were consistent with the histochemical localization of patatin in tubers (Sonnewald et al., 1989), and indicate that the promoter of a class I patatin gene is suitable for expressing glycinin genes in potato tubers. Synthesized normal and modified proteins were processed into pro-forms but not into mature-forms, and accumulated as trimers just like in the ER of soybean cells. Proteins like glycinin generally bear targeting signal to the vacuole in the molecules, which can be recognized in tobacco leaf cells (Utsumi et al., 1993c). Therefore, the proteins expressed in the tuber cells seem likely to be present in the vacuoles as the location corresponding to the protein bodies in the soybean seed and the vacuoles do not contain enzymes which function similarly to the processing enzyme responsible for post-translational cleavage of proglycinin in the protein bodies of soybean seeds (Scott et al., 1992; Muramatsu and Fukazawa, 1993; Shimada et al., 1994). Since the modified proglycinins IV+4Met and V+4Met expressed in *E. coli* exhibit better food functions than the native glycinin from soybean seeds as described in previous section, processing of proglycinin to mature form is not necessary for creating potato tubers producing proteins with good food functions. Efforts to increase the accumulation level are, however, desirable. We are now attempting by three methods: (1) to use genomic DNA of glycinin instead of its cDNA, (2) to co-introduce a gene encoding the enzyme responsible for post-translational processing of proglycinin in the anticipation that the processing of proglycinin to mature form might be essential for the high level accumulation, and (3) to co-introduce a gene for anti-sence RNA of patatin to decrease the synthesis of patatin and increase that of proglycinin concomitantly.

FUTURE PROSPECTS

Our research clearly demonstrates that molecular design of novel soybean glycinins with enhanced food qualities and development of crops producing such glycinins are possible. Determination of three dimensional structure of glycinin and clarification of relationships between the structure and the functional properties at the molecular level are essential for more definite, theoretical, and dramatic improvements. To attain this, X-ray crystallographic analysis of the normal and the modified proglycinins with properties different from those of the native glycinin and the normal proglycinin is required.

REFERENCES

Argos, P., Narayana, S.V.L., and Nelson, N.C., 1985, Structural similarity between legumin and vicilin storage proteins from legumes, *EMBO J.* 4:1111.

Badley, R.A., Atkinson, D., Hauser, H., Oldani, D., Green, J.P., and Stubbs, J.M., 1975, The structure, physical and chemical properties of the soy bean protein glycinin, *Biochim. Biophys. Acta.* 412:214.

Barton, K.A., Thompson, J.F., Madison, J.T., Rosenthal, R., Jarvis, N.P., and Beachy, R.N., 1982, The biosynthesis and processing of high molecular weight precursors of soybean glycinin subunits, *J. Biol. Chem.* 257:6089.

Beachy, R.N., Chen, Z.-L., Horsch, R.B., Rogers, S.G., Hoffmann, N.J., and Fraley, R.T., 1985, Accumulation and assembly of soybean β-conglycinin in seeds of transformed petunia plants, *EMBO J.* 4:3047.

Bogue, M.A., Vonder Haar, R.A., Nuccio, M.L., Griffing, L.R., and Thomas, T.L., 1990, Developmentally regulated expression of a sunflower 11S seed protein gene in transgenic tobacco, *Mol. Gen. Genet.* 222:49.

Chrispeels, M.J., Higgins, T.J.V., and Spencer, D., 1982, Assembly of storage protein oligomers in the endoplasmic reticulum and processing of the polypeptides in the protein bodies of developing pea cotyledons, *J. Cell Biol.* 93:306.

Derbyshire, E., Wright, D.J., and Boulter, D., 1976, Legumin and vicilin, storage proteins of legume seeds, *Phytochemistry* 15:3.

Gidamis, A.B., Mikami, B., Katsube, T., Utsumi, S., and Kito, M., 1994, Crystallization and preliminary X-ray analysis of soybean proglycinins modified by protein engineering, *Biosci. Biotech. Biochem.* 58:703.

Gidamis, A.B., Wright, P., Haque, Z.U., Katsube, T., Kito, M., and Utsumi, S., 1995, Modification tolerability of soybean proglycinin, *Biosci. Biotech. Biochem.* 59:1593.

Hara-Nishimura, I., Inoue, K., and Nishimura, M., 1991, A unique vacuole processing enzyme responsible for conversion of several proprotein precursors into the mature forms, *FEBS Lett.* 294:89.

Higgins, T.J.V., Newbigin, E.J., Spencer, D., Llewellyn, D.J., and Craig, S., 1988, The sequence of a pea vicilin gene and its expression in transgenic tobacco plants, *Plant Mol. Biol.* 11:683.

Kato, A., and Yutani, K., 1988, Correlation of surface properties with conformational stabilities of wild-type and six mutant tryptophan synthase α-subunits substituted at the same position, *Protein Eng.* 2:153.

Katsube, T., Gidamis, A.B., Kanamori, J., Kang, I.J., Utsumi, S., and Kito, M., 1994, Modification tolerability of the hypervariable region of soybean proglycinin, *J. Agric. Food Chem.* 42:2649.

Kim, C.-S., Kamiya, S., Kanamori, J., Utsumi, S., and Kito, M., 1990a, High-level expression, purification and functional properties of soybean proglycinin from *Escherichia coli*, *Agric. Biol. Chem.* 54:1543.

Kim, C.-S, Kamiya, S., Sato, T., Utsumi, S., and Kito, M., 1990b, Improvement of nutritional value and functional properties of soybean glycinin by protein engineering, *Protein Eng.* 3:725.

Kinsella, J.E., 1979, Functional properties of soy proteins, *J. Am. Oil Chem. Soc.* 56:242.

Kito, M., Moriyama, T., Kimura, Y., and Kambara, H., 1993, Changes in plasma lipid levels in young heathy volunteers by adding an extruder-cooked soy protein to conventional meals, *Biosci. Biotech. Biochem.* 57:354.

Miki, T., Yasukochi, T., Nagatani, H., Furuno, M., Orita, T., Yamada, H., Imoto, T., and Horiuchi, T., 1987, Construction of a plasmid vector for the regulatable high level expression of eukaryotic genes in *Escherichia coli* : An application to overproduction of chicken lysozyme, *Protein Eng.* 1:327.

Millerd, A., 1975, Biochemistry of legume seed proteins, *Annu. Rev. Plant Physiol.* 26:53.

Mori, T., Utsumi, S., and Inaba, H., 1979, Interaction involving disulfide bridges between subunits of soybean seed globulin and between subunits of soybean and sesame seed globulins, *Agric. Biol. Chem.* 43:2317.

Mori, T., Nakamura, T., and Utsumi, S., 1982, Gelation mechanism of soybean 11S globulin: Formation of soluble aggregates as transient intermediates, *J. Food Sci.* 47:26.

Muramatsu, M., and Fukazawa, C., 1993, A high-order structure of plant storage proprotein allows its second conversion by an asparagine-specific cysteine protease, a novel proteolytic enzyme, *Eur. J. Biochem.* 215:123.

Nakamura, T., Utsumi, S., Kitamura, K., Harada, K., and Mori, T., 1984, Cultivar differences in gelling characteristics of soybean glycinin, *J. Agric. Food Chem.* 32:647.

Nielsen, N.C., 1984, The chemistry of legume storage proteins, *Philos. Trans. R. Soc. London, Ser. B* 304:287.

Pikaard, C.S., Brusca, J.S., Hannapel, D.J., and Park, W.D., 1987, The two classes of genes for the major potato tuber protein, patatin, are differentially expressed in tubers and roots, *Nucleic Acids Res.* 15:1979.

Racusen, D., and Foote, M., 1980, A major soluble glycoprotein of potato tubers, *J. Food Biochem.* 4:43.

Robert, L.S., Thompson, R.D., and Flavell, R.B., 1989, Tissue-specific expression of a wheat high molecular weight glutenin gene in transgenic tobacco, *Plant Cell* 1:569.

Saalbach, G., Jung, R., Kunze, G., Manteuffel, R., Saalbach, I., and Muntz, K., 1990, Expression of modified legume storage protein genes in different systems and studies on intracellular targeting of *Vicia faba* legumin in yeast, in: *Genetic Engineering of Crop Plants*, G.W. Lycett and D. Grierson, eds., Butterworth, London.

Schoemaker, J.M., Brasnett, A.H., and Marston, F.A.O., 1985, Examination of calf prochymosin-containing inclusion bodies, *EMBO J.* 4:775.

Scott, M.P., Jung, R., Muntz, K., and Nielsen, N.C., 1992, A protease responsible for post-translational cleavage of a conserved Asn-Gly linkage in glycinin, the major seed storage protein of soybean, *Proc. Natl. Acad. Sci. U.S.A.* 89:658.

Sengupta-Gopalan, C., Reichert, N.A., Barker, R.F., and Hall, T.C., 1986, Developmentally regulated expression of the bean β-phaseolin gene in tobacco seed, *Proc. Natl. Acad. Sci. U.S.A.* 82:3320

Shimada, T., Hiraiwa, N., Nishimura, M., and Hara-Nishimura, I., 1994, Vacuolar processing enzyme of soybean that converts proproteins to the corresponding mature forms, *Plant Cell Physiol.* 35:713.

Sonnewald, U., Studer, D., Rocha-Sosa, M., and Willmitzer, L., 1989, Immunocytochemical localization of patatin, the major glycoprotein in potato *(Solanum tuberosum* L.) tubers, *Planta* 178:176.

Staswick, P.E., Hermodson, M.A., and Nielsen, N.C., 1981, Identification of the acidic and basic subunit complexes of glycinin, *J. Biol. Chem.* 256:8752.

Staswick, P.E., Hermodson, M.A., and Nielsen, N.C., 1984, Identification of the cysteines which link the acidic and basic components of the glycinin subunits, *J. Biol. Chem.* 259:13431.

Takaiwa, F., Oono, K., Wing, D., and Kato, A., 1991, Sequence of three members and expression of a new major subfamily of glutelin genes from rice, *Plant Mol. Biol.* 17:875.

Takaiwa, F., Katsube, T., Kitagawa, S., Higasa, T., Kito, M., and Utsumi S., 1995, High level accumulation of soybean glycinin in vacuole-derived protein bodies in the endosperm tissue of transgenic tobacco seed, *Plant Sci.* 111:39.

Tumer, N.E., Richter, J.D., and Nielsen, N.C., 1982, Structural characterization of the glycinin precursors, *J. Biol. Chem.* 257:4016.

Uhlin, B.E., Molin, S., Gulstafsson, P., and Nordstrom, K., 1979, Plasmid with temperature-dependent copy number for amplification of cloned gene and their products, *Gene* 6:91.

Utsumi, S., 1992, Plant food protein engineering, *Adv. Food Nutr. Res.* 36:89.

Utsumi, S., and Kito, M., 1991, Improvement of food protein functions by chemical, physical, and biological modifications, *Comments Agric. Food Chem.* 2:261.

Utsumi, S., Kim, C.-S., Kohno, M., and Kito, M., 1987, Polymorphism and expression of cDNAs encoding glycinin subunits, *Agric. Biol. Chem.* 51:3267.

Utsumi, S., Kim, C.-S., Sato, T., and Kito, M., 1988a, Signal sequence of preproglycinin affects production of the expressed protein in *Escherichia coli*, *Gene* 71:349.

Utsumi, S., Sato, T., Kim, C.-S., and Kito, M., 1988b, Processing of preproglycinin expressed from cDNA-encoding AlaB1b subunit in *Saccharomyces cerevisiae*, *FEBS Lett.* 233:273.

Utsumi, S., Kanamori, J., Kim, C.-S., Sato, T., and Kito, M., 1991, Properties and distribution of soybean proglycinin expressed in *Saccharomyces cerevisiae*, *J. Agric. Food Chem.* 39:1179.

Utsumi, S., Gidamis, A.B., Kanamori, J., Kang, I.J., and Kito, M., 1993a, Effects of deletion of disulfide bonds by protein engineering on the conformation and functional properties of soybean proglycinin, *J. Agric. Food Chem.* 41:687.

Utsumi, S., Gidamis, A.B., Mikami, B., and Kito, M., 1993b, Crystallization and preliminary X-ray crystallographic analysis of the soybean proglycinin expressed in *Escherichia coli*, *J. Mol. Biol.* 233:177.

Utsumi, S., Kitagawa, S., Katsube, T., Kang. I.J., Gidamis, A.B., Takaiwa, F., and Kito, M., 1993c, Synthesis, processing and accumulation of modified glycinins of soybean in the seeds, leaves and stems of transgenic tobacco, *Plant Sci.* 92:191.

Utsumi, S., Kitagawa, S., Katsube, T., Higasa, T., Kito, M., Takaiwa, F., and Ishige, T., 1994, Expression and accumulation of normal and modified soybean glycinins in potato tubers, *Plant Sci.* 102:181.

Utsumi, S., Matsumura, Y., and Mori, T., Structure-functionality relationship of soy proteins, in :*Food Proteins and Their Applications*, S. Damodaran and A. Paraf, eds., Marcel Dekker, New York, in press.

von der Geest, A.H.M., Hall, G.E., Spiker, S., and Hall. T.C., 1994, The β-phaseolin gene is flanked by matrix attachment regions, *Plant J.* 6:413.

Williams, D.C., Van Frank, R.M., Muth, W.L., and Burnett, J.P., 1982, Cytoplasmic inclusion bodies in *Escherichia coli* producing biosynthetic human insulin proteins, *Science* 215:687.

Williamson, J.D., Galili, G., Larkins, B.A., and Gelvin, S.B., 1988, The synthesis of a 19 kilodalton zein protein in transgenic Petunia plants, *Plant Physiol.* 88:1002.

Wright, D.J., 1988, The seed globulins, in: *Developments in Food Proteins-6*, B.J.F. Hudson, ed., Elsevier, London.

HIGH FRUCTOSE SYRUPS: EVALUATION OF A NEW
GLUCOSE ISOMERASE FROM <u>STREPTOMYCES</u> SP.

Todd W. Gusek,[1] K. Sailaja,[2] and Richard Joseph[3]

[1]Cargill, Inc.
2301 Crosby Road
Wayzata, MN 55391

[2] Indian Institute of Science
Molecular Biophysics Unit
Bangalore - 560 012
India

[3] Central Food Technological Research Institute
Department of Microbiology
Mysore - 570 013
India

INTRODUCTION

High Fructose Corn Syrup Market

The commercial success of high fructose corn syrup (HFCS) is remarkable. The first shipment of enzymatically produced fructose corn syrup in the United States was made by the Clinton Corn Processing Co. in 1967. Early products contained 15% fructose and were prepared using soluble glucose isomerases, at high production costs (Guzman-Maldonado and Paredes-Lopez, 1995). Within twenty years U.S. production had grown to over 11 billion pounds dry solids (Long, 1991). Clinton Corn Processing Co. received the prestigious International Food Technology Industrial Achievement Award in 1975 for its pioneering efforts in developing the immobilized enzyme technology for economical production of HFCS.

Glucose isomerase is the enzyme which catalyzes the conversion of glucose to fructose. Commercial interest in glucose isomerase centers on the gain in sweetness that is achieved when a glucose syrup is treated with the enzyme to attain a fructose-enriched product. Glucose has roughly 70% of the sweetness of sucrose, whereas fructose is up to 60% sweeter than sucrose depending on the application (Antrim et al. 1979). The reaction mixture at equilibrium is equisweet with sucrose and contains approximately 42% fructose, 52% dextrose, and 6% higher saccharides. The incentive to use high fructose syrups in the United States is predominantly economic, considering the average price differential between refined sucrose and HFCS. In 1993, U.S. wholesale prices for refined sugar were 25.2 cents per pound, compared with the two main fructose syrup products, 42% HFCS (18.8 ¢/lb) and 55% HFCS (20.9 ¢/lb) (Mancini, 1994). Additional driving forces include the availability of inexpensive sources of starchy raw materials (corn), synergistic sweetening effects between fructose and other sweeteners (both nutritive and non-nutritive), and inhibition of sugar crystallization in finished products. HFCS is a fluid so it can be pumped, which offers a processing advantage over the handling of bulk sugar.

In 1972, 26 billion pounds of nutritive sweeteners were consumed in the United States. Refined sugar (sucrose) comprised 82% of this total while HFCS accounted for a mere 1%. Over the next twenty years, nutritive sweetener consumption grew by an annual average of 2%, to just under 37 billion pounds in 1992 (144 pounds per capita; 1995 Corn Annual). Sugar dropped to 45% of the total volume in 1992, while HFCS jumped to 36% (Mancini, 1994) (Figure 1). U.S. sales of HFCS were $2.97 billion in 1991, with 98% going to domestic usage. The soft drink industry is by far the largest customer of HFCS. It consumed 70% of HFCS volume in 1993. Bakery, canned, and dairy products utilized the remainder. The 55 HFCS product accounts for about 65% of HFCS volume (Long, 1991).

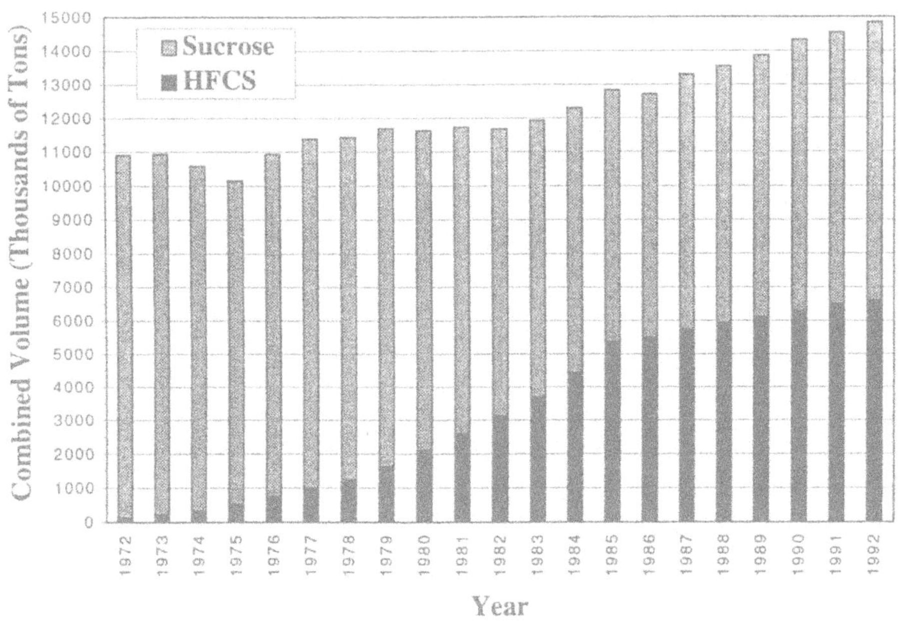

Figure 1. Growth of high fructose corn syrup market in the United States, between 1972-92. Sucrose and HFCS represented a combined volume of 83% of nutritive sweetener consumption in 1972, and 81% in 1992. (Data from: Mancini, 1994.)

HFCS Production

The multiple processing steps required to convert corn starch into dextrose syrup are summarized in Figure 2. The isomerization of D-glucose (dextrose) to fructose is conducted using a reactor which incorporates glucose isomerase fixed to a water insoluble resin. The key manufacturing steps are listed in Figure 3. The significance of glucose isomerase is reflected in the fact that it represents one of the three highest value and highest tonnage industrial enzymes worldwide, alongside amyloglucosidase (glucoamylase) and bacterial protease. Glucose isomerase also represents the largest application of immobilized enzyme technology. Immobilization of enzymes provides a number of advantages including continuous reactor operation, enzyme reusability, and simplified product purification. Each feature translates to reduced production costs.

D-glucose isomerase is synonymously used for D-xylose isomerase (D-xylose ketol-isomerase, EC 5.3.1.5), an enzyme that catalyzes the reversible isomerization of D-xylose into D-xylulose. A number of bacteria produce the enzyme intracellularly, including the genera <u>Actinoplanes</u>, <u>Arthrobacter</u>, <u>Bacillus</u>, <u>Lactobacillus</u>, and <u>Streptomyces</u>. Glucose isomerases are homotetrameric proteins with molecular weights around 160,000 daltons. They have a temperature optimum near 60°C and pH optimum between 7.5-9. Glucose isomerases require bivalent cations (Mg^{2+}, Mn^{2+}) for activation. Aside from their role in the catalytic mechanism, bivalent cations (e.g., Co^{2+}) also appear to increase thermostability of some glucose isomerases. Calcium (Ca^{2+}), on the other hand, acts as an inhibitor and must be removed from the feed syrup prior to its conversion.

In view of the potential benefits of identifying an enzyme that demonstrates greater thermostability than existing industrial isomerases, or developing a process that is efficient below pH 7.5 where fructose is more stable, a novel glucose isomerase from <u>Streptomyces coelicolor</u> was immobilized and evaluated. Reported here are two immobilization schemes, and the results of applying the preferred isomerase conjugate to semi-continuous production of high fructose syrup.

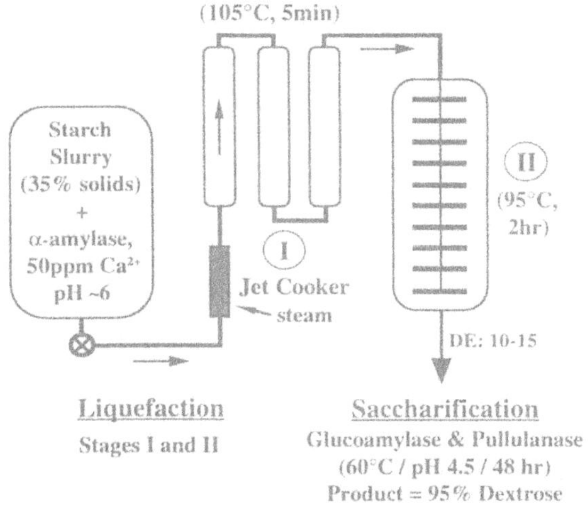

Figure 2. Conversion of starch by liquefaction and saccharification. Efficient liquefaction employs a jet-cooking process whereby suspended granular starch is first gelatinized using a combination of high temperature and shear (I), and then dextrinized in a stirred-tank reactor (II) (Olsen, 1993). Thermostable bacterial amylase is used to randomly hydrolyze endo- α-1,4-glycosidic linkages and give a product with dextrose equivalent (DE) in the vicinity of 10-15. Saccharification of the maltodextrins is achieved by further hydrolysis using a debranching (α-1,6) enzyme (bacterial pullulanase) together with fungal glucoamylase (an exo-amylase). High contents of glucose (95-97%) can be produced from most starch raw materials (corn, wheat, potatoes, tapioca, barley, and rice).

MATERIALS AND METHODS

Production of Glucose Isomerase

<u>Streptomyces</u> <u>coelicolor</u> strain A3 (2) ("coelicolor" is latin for sky color; refers to blue pigments the bacterium elaborates during growth on solid media) was cultivated at 30°C in a medium that contained starch and peptone as the principal carbon and nitrogen sources,

respectively (Sailaja and Joseph, 1993). After 16h of growth, D-xylose was added to the shake flasks to induce intracellular production of the xylose (glucose) isomerase. Mycelium was harvested after an additional 32h of growth and treated with lysozyme to liberate the enzyme. Soluble isomerase was not further purified for the immobilization studies.

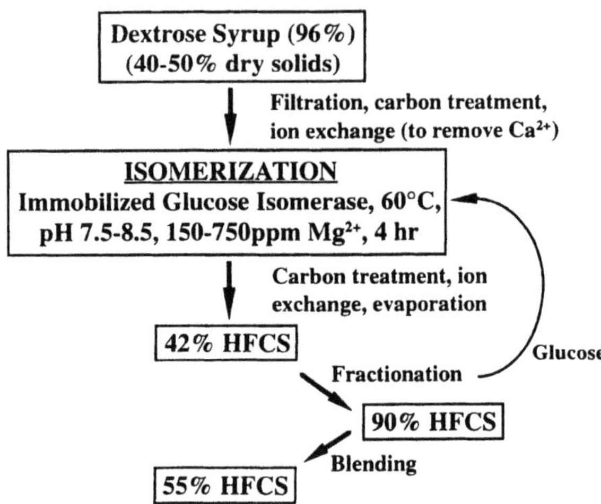

Figure 3. Flow chart for the manufacture of 42, 55 and 90% high fructose corn syrup. The dry substance content of the dextrose feed syrup is kept below 50% to maintain efficient diffusion of dextrose (substrate) to, and fructose (product) away from the immobilized enzyme. Dissolved oxygen can be removed by de-aeration to reduce formation of byproducts. Magnesium ions activate glucose isomerase (GI), and also competitively inhibit the action of residual calcium ions, which are potent inhibitors of GI. Reactors typically have a diameter between 2-5 ft and bed height between 6-16 ft. Good flow distribution is ensured by maintaining a minimum bed height to diameter ratio of 3:1. During operation, immobilized GI loses activity. A reactor load of GI is replaced after three half-lives, or when its activity has dropped to approximately 12% of its initial value. The most stable isomerases have half-lives of 200 days in commercial operation (Olsen, 1993). The equilibrium conversion of glucose to fructose is 50% under industrial conditions, but is normally limited to 45% to minimize sugar degradation and development of color. A syrup highly enriched in fructose is made by pumping the 42% stream into a fractionation unit which employs calcium sites to retain the fructose while dextrose and the higher saccharides pass through. The resultant 80-90% fructose syrup is refined and blended with 42% HFCS to make the 55% product (Long, 1991). (Figure adapted from Guzman-Maldonado and Paredes-Lopez, 1995.)

Immobilization of Glucose Isomerase to Granular DEAE Cellulose

Granular DEAE (diethylaminoethyl) cellulose, or GDC, was kindly supplied by Dr. Richard Antrim (Genencor International, Cedar Rapids, IA). GDC is a composite material formed by mixing one part chemical grade cellulose with one part alumina (or another densification agent, typically a ceramic oxide) and compounding the mixture with two parts polystyrene (Antrim and Hurst, 1982). The polystyrene provides dimensional support to minimize channeling and development of packing problems that can occur in deep bed reactors. The cellulose-alumina-polystyrene agglomerate is ground and sized to 40-100 mesh, and derivatized with diethylaminoethyl-chloride hydrochloride, to form GDC. GDC has a high adsorptive capacity, between 1200-1800 IGIU/g dry resin. IGIU, or international glucose isomerase unit, is the amount of enzyme which will convert 1 μmole glucose to fructose per minute in a solution that initially contains 2M glucose, 20mM $MgSO_4$ and 1mM $CoCl_2$ at pH 6.84 (0.2M sodium maleate) and temperature 60°C. Purified glucose isomerase has a specific activity of 40 IGIU/mg (R. Antrim, personal communication). Therefore one gram of GDC should bind approximately 40mg pure glucose isomerase.

The resin was first regenerated by repeated washing with 0.5N HCl followed by 0.5N NaOH. It was then flushed with water and equilibrated with 50mM sodium phosphate, pH 7.0. GDC was loaded into a small column. Freshly prepared, unpurified S. coelicolor cell lysate containing 0.5 units glucose isomerase activity per ml was added to the GDC support. (One unit is defined as the amount of enzyme which catalyzes the formation of one μmole D-fructose or D-xylulose per minute under the assay conditions below.) The eluate was monitored for glucose- and xylose isomerase activities to determine the extent of enzyme adsorption.

Immobilization of Glucose Isomerase to a Ceramic Carrier

A second pathway for immobilization was identified for direct comparison of enzyme stability and reactor performance to the GDC conjugate. Because GDC is an organic support material onto which enzymes are immobilized by electrostatic adsorption, it was decided that the alternative method should utilize an inorganic carrier and employ covalent attachment of glucose isomerase (Figure 4). Aluminum oxide (alumina; Al_2O_3) is an inexpensive ceramic support that is used for both covalent coupling and physical adsorption of proteins to its surface. Controlled pore alumina (neutral, Brockmann I; 150 mesh, mean pore diameter 58Å, surface area 155 m²/g) was obtained from Aldrich Chemical Co., Inc. (Milwaukee, WI). The alumina was derivatized by aqueous silanization (procedure of Eby and Schuerch, 1975) and then activated with glutaraldehyde (Stolzenbach and Kaplan, 1976). γ-amino-propyltriethoxysilane was obtained from Aldrich, and glutaraldehyde (25%) from Sigma Chemical Co. (St. Louis, MO). S. coelicolor lysate containing 0.5 units/ml GI activity was slowly loaded into a small glass column prepacked with activated alumina. Unbound proteins were removed by flushing the column with 50mM phosphate buffer pH 7.0.

Assays

Glucose and xylose isomerase activities were measured by the method of Suekane et al. (1978). The substrate solution contained either 1M D-glucose (dextrose) or 0.25M xylose, and both 10mM $MgSO_4$ and 1mM $CoCl_2$. It was buffered at pH 7.0 with 0.2M sodium phosphate. After 30min incubation at 60°C, the reaction was halted by addition of an equal volume of 0.5M perchloric acid. Ketose sugar was determined by the cysteine-carbazole colorimetric method of Dische and Borenfreund (1951). Activity of immobilized glucose isomerase was initially evaluated in small-scale glass column reactors, at 30°C over a range of flow rates. The dextrose substrate solution was the same as above, except that it was buffered with 50mM sodium phosphate. Glucose to fructose conversion was again monitored by the cysteine-carbazole assay. A jacketed reactor connected to a circulating water bath was used for semi-continuous operation over a range of temperatures. When the columns were not in use, the resins were kept in phosphate buffer that contained a trace of sodium azide, at ambient temperature.

RESULTS AND DISCUSSION

Binding Efficiency

The GDC carrier demonstrated considerably greater binding capacity for proteins in general, and glucose isomerase in particular, than the alumina (Table 1). This result reflects the high loading capacity of fibrous ion exchange cellulosic preparations. GDC was designed with this feature because high isomerase loads increase glucose conversion rates, thereby decreasing the period during which fructose is held under conditions that promote formation of undesirable byproducts.

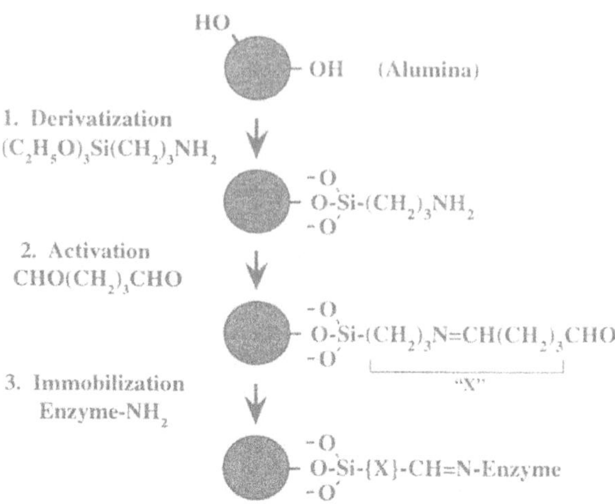

Figure 4. Sequential steps for modifying the surface chemistry of alumina, to provide a support for enzyme immobilization. The starting material, also called "active" alumina, is obtained by controlled heating of hydrated aluminum oxide powders to eliminate their water of constitution. The alumina is first derivatized with γ-aminopropyltriethoxysilane. Activation of alkylamine alumina requires one of the two aldehydic groups of glutar[di]aldehyde. The second group is available for covalent coupling to a free ε-amino moiety (from lysine) on the enzyme. The $\underline{S}$. $\underline{coelicolor}$ glucose isomerase has three lysine residues (Sailaja and Joseph, 1993). The amino-carbonyl reaction forms an imine (Schiff base).

Table 1. Binding efficiency of $\underline{S}$. $\underline{coelicolor}$ glucose isomerase to two resins.

	Granular DEAE Cellulose (GDC)	Alumina
Protein Content in Lysate: % Bound	85[1]	8
Glucose Isomerase Activity Units (Lysate): % Bound	100	5
Apparent Binding Capacity (mg protein / gram resin)	>24	2

[1]It is unreasonable to expect 100% because protein binding is by electrostatic adsorption, which requires that the ligand be anionic to exchange ions with the DEAE support.

Semi-Continuous Conversion of Glucose to Fructose

In consideration of the superior performance of GDC for immobilization of the $\underline{S}$. $\underline{coelicolor}$ glucose isomerase, a jacketed reactor with an 8:1 dimensional ratio between its bed height and diameter was loaded with 25g (dry weight) GDC-isomerase conjugate that contained approximately 8 units of GI activity. Enzyme-mediated isomerization of glucose to fructose was clearly temperature-dependent (Figure 5). The highest initial rate of conversion at a given substrate residence time occurred around 70°C (data not shown). An accurate determination was not possible because of difficulty in maintaining constant temperature above 65°C. This result is consistent with analysis of the soluble enzyme, which demonstrated an apparent temperature optimum for activity at 70°C. Prolonged operation (3-4 days) of the reactor above 65°C, however, resulted in >50% loss of original activity. Commercial glucose isomerases exhibit thermal instability at temperatures around 70°C. The same was observed for the S. coelicolor isomerase.

The equilibrium constant for glucose isomerization is close to unity, so under optimal process conditions about 50% of the glucose should be converted. Industrial processes are designed to provide around 45% conversion. A maximum of 23% conversion of glucose to fructose was observed in the laboratory trials. Two key factors influence the efficiency of an immobilized enzyme reactor: (a) substrate residence time, and (b) amount of enzyme per unit of resin. Suboptimal values for each factor would account for reduced glucose conversion.

Residence time is inversely proportional to the flow rate of substrate through the reactor. Fructose is somewhat unstable at conventional conversion temperatures and alkaline pH; therefore commercial reactors use high enzyme loadings to minimize the residence time required to achieve 45% conversion. Assuming comparable specific activities for the S. coelicolor glucose isomerase and that used by Antrim and Hurst (1982) in trials with GDC, the eight units of isomerase that were immobilized to GDC in the present study represented less than 0.02% of the adsorptive capacity of the resin! This load is negligible compared to standard industrial practice. A followup test using partially purified glucose isomerase would give significantly higher enzyme loadings and increase the isomerization potency of the GDC reactor.

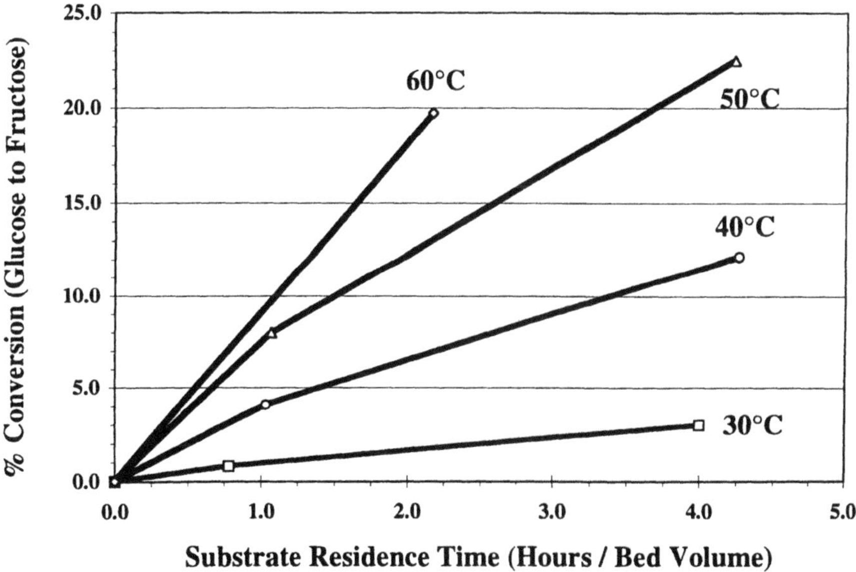

Figure 5. Influence of temperature on rate of conversion of glucose to fructose in an immobilized enzyme column reactor that contained S. coelicolor glucose isomerase bound to granular DEAE cellulose. Substrate (pH 7.0) contained 1M glucose, 10mM $MgSO_4$, 1mM $CoCl_2$, and 50mM sodium phosphate. Residence time was dictated by operational flow rate, and refers to the length of time that a volumetric unit of substrate spent in the vicinity of the isomerase-GDC conjugate.

Stability of Immobilized Isomerase During Storage

In spite of its marginal capacity for protein loading, alumina demonstrated significantly greater retention of immobilized enzyme activity during prolonged storage in buffer at ambient temperature, compared to isomerase coupled to GDC. After 80 days, the isomerase-alumina conjugate exhibited 91% of the original isomerization potency at a given flow rate. The GDC-isomerase conjugate fell to 35% during the same period. The rate of activity loss over the 80-day period was essentially linear for alumina. The rate of glucose isomerase activity loss on GDC was disproportionately high during the first 30 days, but levelled off after 50-60 days.

The disparity in activity loss probably reflects the different linkages between ligand and the two carriers. Proteins that are covalently coupled to their carrier, as in the case of alumina, are expected to experience considerably less leaching or migration between binding sites, than are proteins which are adsorbed to a resin. The relative strength of electrostatic forces between a protein and ionic resin is influenced by the pH and ionic strength of the solvent.

It is worth noting that the ionic nature of ligand adsorption to GDC simplifies regeneration of the reactor, and at low cost. Extended operation of an immobilized glucose isomerase reactor results in enzyme denaturation, and the accumulation of substrate debris and other proteinaceous materials collectively diminish the potency of the reactor. GDC binding sites are readily refreshed by treating the resin with a solution of alkali. A solution of soluble enzyme can then be brought into contact with the carrier through ion exchange, to restore the original glucose conversion capacity.

SUMMARY

Glucose isomerase from <u>Streptomyces</u> <u>coelicolor</u> was immobilized to activated alumina and granular DEAE cellulose. Each conjugate converted glucose to fructose. GDC proved to be the preferred resin because the immobilization procedure (involving ion exchange) was simple and yielded higher enzyme loadings and increased reactor efficiency. Column reactors incorporating either resin operated satisfactorily over a range of temperatures and flow rates. A syrup containing 23% fructose was prepared using the GDC reactor. Increased conversion of glucose to the industrial target of 45% is possible by exploiting the high loading capacity of fibrous cellulose.

Immobilization to activated alumina presumably involved the formation of covalent bonds between enzyme and carrier. It was anticipated that a covalently-coupled ligand would experience little to no leaching from the support. The data demonstrated that alumina provided a more stable conjugate than GDC (as measured by retention of isomerase activity during prolonged storage). This apparent benefit of alumina is overshadowed by the high enzyme capacity and relative ease of preparing and regenerating GDC-based reactors. These features translate to reduced processing costs. In fact, GDC has become the industrial resin of choice for production of high fructose syrup.

FUTURE RESEARCH

The pH profile for activity of the <u>S</u>. <u>coelicolor</u> glucose isomerase (optimum ≈ 7.0) is among the lowest reported to date. Greater than 60% of the maximum activity is still present at pH 6.0. HFCS manufacturers would find immediate benefits using isomerases with lower pH requirements for activity. Isomerization under alkaline conditions favors formation of non-dextrose and non-fructose degradation products, which contribute to reduced product sweetness and development of color and off-flavors. These reactions are retarded at lower pH. Furthermore, less alkali should be required to neutralize the saccharified liquor (pH 4.5) from the previous step, if the pH for commercial isomerization is shifted down by ≥0.5 units (currently between 7.5-7.8). Enzyme producers are working to develop isomerases with reduced pH optima by genetic manipulation (Lambeir et al. 1994). This enzyme may already provide the desired property. Current research is directed toward overexpression of the glucose isomerase gene to increase production levels.

ACKNOWLEDGEMENTS

The J. William Fulbright Foreign Scholarship Board and the Council for International Exchange of Scholars are acknowledged for awarding TWG a Fulbright Research Grant. The Central Food Technological Research Institute, Mysore, India is recognized for serving as the host institution for the study. Special thanks are extended to the Eastman Kodak Co. for providing photographic film to record the rich cultural and geographical diversity of India observed during periodic assignments in other parts of the country.

REFERENCES

Antrim, R.L., Colilla, W., and Schnyder, B.J., 1979, Glucose isomerase production of high-fructose syrups, *Appl. Biochem. Bioeng.* 2:97.

Antrim, R.L. and Hurst, L.S., 1982, Process for preparing agglomerated fibrous cellulose, United States Patent 4,355,117.

Dische, Z. and Borenfreund, E., 1951, A new spectrophotometric method for the detection and determination of keto sugars and trioses, *J. Biol. Chem.* 192:583.

Eby, R. and Schuerch, C., 1975, Solid-phase synthesis of oligosaccharides, *Carbohydrate Res.* 39:151.

Guzman-Maldonado, H. and Paredes-Lopez, O., 1995, Amylolytic enzymes and products derived from starch: a review, *Crit. Rev. Food Sci. Nutr.* 35:373.

Lambeir, A.M., Lasters, I., Mrabet, N., Quax, W.J., Van der Laan, J.M., and Misset, O., 1994, Modified prokaryotic glucose isomerase enzymes with altered pH activity profiles, United States Patent 5,340,738.

Long, J.E., 1991, High fructose corn syrup, in: *Alternative Sweeteners,* 2nd Edition, L. O'Brien Nabors and R.C. Gelardi, eds., Marcel Dekker, Inc., New York.

Mancini, T., 1994, *The U.S. Sweetener Market,* Business Trend Analysts, Inc., Commack, NY.

1995 Corn Annual, Corn Refiners Association, Inc., Washington, D.C.

Olsen, H.S., 1993, Enzymes for starch modification, *Food Technol. Int. Europe* :121.

Sailaja, K. and Joseph, R., 1993, Purification and properties of xylose isomerase of <u>Streptomyces</u> <u>coelicolor</u> A3 (2), *Starke* 45:306.

Stolzenbach, F.E. and Kaplan, N.O., 1976, Immobilization of lactic dehydrogenase, *Methods Enzymol.* 44:929.

Suekane, M., Tamura, M., and Tomimura, C., 1978, Purification and properties of glucose isomerase from <u>Streptomyces</u> <u>olivochromogenes</u> and <u>Bacillus</u> <u>stearothermophilus</u>, *Agric. Biol. Chem.* 42:909.

SERINE PROTEINASES FROM COLD-ADAPTED ORGANISMS

Magnús M. Kristjánsson, Bjarni Ásgeirsson, and Jón B. Bjarnason

Science Institute,
Department of Chemistry,
University of Iceland,
Reykjavik, Iceland

INTRODUCTION

Proteins and especially enzymes from organisms that have adapted to extreme conditions in environmental temperatures have been a subject of considerable interest in both basic and applied research for number of years. Sofar most of the research has focused on enzymes from thermophilic microorganisms. As enzymes from thermophiles are generally found to be more thermostable than their counterparts from mesophiles, they have received much attention as experimental model systems for studying the underlying molecular principles in thermostabilization of proteins. Despite much research effort , the question of how thermophilic proteins are stabilized to withstand temperatures close to or above the boiling point of water is,however, still unsolved. Because of their high activity and stability at elevated temperatures, enzymes from thermophiles have also been considered an attractive alternative to mesophilic enzymes, that despite of their often limited thermal stability are used in several industrial applications that require high operational temperatures.

Cold is the most widespread physiological stress condition that organisms have to adapt to, or avoid. Adaptive changes in protein structure and function, induced by cold are of prime importance for cold acclimation and survival processes (Franks, 1985, 1995). Enzymes from cold-adapted organisms have nevertheless been much less studied and relatively little is known about mechanisms underlying cold-adaptation of proteins. These enzymes are usually found to have higher specific activities at low temperatures than their counterparts from organisms adapted to higher temperatures, a property that may make these enzymes potential candidates for several industrial applications, such as in the processing of sensitive biological materials, including foodstuffs, that have to be carried out under chilled or refrigerated conditions. Enzymes from cold-adapted organisms are also often found to be comparatively thermolabile, which can be beneficial in operations when enzyme treatment has to be terminated rapidly, without excessive heat treatment of the raw material.

The serine proteinases are one of the best characterized groups of enzymes. They are traditionally classified into two families; the chymotrypsin and the subtilisin families, that share

Food Proteins and Lipids
Edited by Damodaran, Plenum Press, New York, 1997

a common catalytic site, but are otherwise structurally different. The best known representatives of the chymotrypsin family are the digestive proteinases, trypsin, chymotrypsin and elastase, but other members of this family are involved in various cellular functions, including blood clotting, fibrinolysis, complement activation, fertilization and in the production of hormones and pharmacologically active peptides (Neurath, 1984, Bond and Butler, 1987). These enzymes have been isolated and characterized from several different species from bacteria to humans and crystal structures are available of several of them. The subtilisin family (subtilases) is also an extensively studied group of enzymes. Earlier it was thought that subtilases were only found in prokaryotes, but now it has become clear that these enzymes are widely distributed in biological systems, both in lower as well as in higher eukaryotes, including humans. The subtilases have been subdivided into two major classes on basis of sequence homologies (Siezen et al, 1991). Subtilases of class I are represented by the "true" subtilisins from *Bacilli*, but also include enzymes from other bacteria, lower eukaryotes and mammals. The best known member of class II subtilases is proteinase K from the fungi *Tritirachium album* Limber, but this class also includes related fungal proteinases and enzymes from yeast and bacteria (Siezen et al., 1991). High resoluton crystal strucures are available for the subtilisins (Bode et al., 1987, Bott et al., 1988, McPhalen and James, 1988, Betzel et al., 1992), including the thermophilic thermitase (Teplyakov et al., 1990), and for proteinase K (Betzel et al., 1990, Müller et al., 1994). The subtilisins have been studied extensively by protein engineering techniques and a multitude of mutants of the enzymes have been generated for the purpose of studying their catalytic properties, stability and structure (Pantoliano et al., 1988, Wells and Estell,1988, Carter and Wells, 1990, Braxton and Wells, 1992, Strausberg et al., 1993, Rheinecker et al., 1994, Pedersen et al., 1994).

Much of the longstanding interest in the subtilisins stems from their importance as industrial enzymes, where their primary use is as additives in laundry detergents. As the trend has been towards lower washing temperatures, because of energy saving considerations, serine proteinases from cold-adapted organisms may be an interesting alternative to the present detergent proteinases in some applications. Alternatively, an understanding of the underlying principles of cold-adaptation of these proteinases may provide strategies to engineer existing detergent enzymes with higher activites at low temperatures.

In this paper we will attempt to summarize some of the available data on cold-adaptive properties of serine proteinases from organisms adapted to low temperatures. We will view the information on low temperature adaptation from the perspective of our own studies of enzymes from the Atlantic cod and reflect on some recent studies of proteinases from psychrophilic microorganism.

TEMPERATURE ADAPTATION

Studies on proteins from thermophiles and mesophiles show that there apparently exist no general rules in thermostabilization of proteins, but rather that enhancement of thermal stability may be achieved by a cooperation of several small changes in noncovalent interactions over the entire protein molecule (Zuber, 1988, Jaenicke, 1991, Adams et al., 1995). Attempts have been made to correlate thermal stability to certain amino acid exchanges found to occur in comparative studies on amino acid sequences of homologous mesophilic and thermophilic proteins (Argos et al.,1979, Menendez-Arias and Argos, 1989). While sets of rules obtained in such statistical sequence comparisons may provide some guidelines for engineering increased thermal stability into mesophilic proteins, the data presently available does not indicate that general rules for "hot" amino acid exchanges exist that can explain thermophilic behaviour (Jaenicke, 1991, Böhm and Jaenicke, 1994, Korndörfer et al., 1995). Available crystal structures of hyperthermophilic proteins have shown that these structures are very similar to the enzymes from mesophiles despite large differences in stability (Day et al., 1992, Korndörfer et al., 1995, Chan et al., 1995). Only minimal structural alterations involving extra surface salt-brigdes, improved hydrogen bonding or more favorable localized hydrophobic contacts, may account for the high degree of thermostabilization observed in these proteins (Blake et al., 1992, Day et al., 1992, Korndörfer et al., 1995). Free energies of stabilization of native protein structures are marginal, being typically of the order 20-80 kJ mol^{-1}, or what amounts to the energy of few hydrogen bonds, salt-brigdes or hydrophobic interactions. This net stability is achieved by a delicate balance between large stabilizing and destabilizing forces, but as a result of this delicate balance only minor shifts in contributions of these forces, such as by small improvements of several interactions at different locations within the protein molecule, can significantly alter the stability of a protein (Jaenicke and Zavodsky, 1990, Jaenicke, 1991). Although the physical basis for the stability of the folded state of proteins is still not fully understood, hydrophobic interactions have been emphasized as the major stabilizing force in proteins (Dill, 1990, Doig and Williams, 1991), despite speculations to the contrary (Murphy et al., 1990). A recent study in which thermodynamic data for the stability of several mesophilic globular proteins was evaluated, came up with a model that suggested that hydrophobic interactions would not stabilize proteins having melting temperatures of about 87 °C or above (Ragone and Colonna, 1995). Other forces such as hydrogen bonds and van der Waals interactions would therefore be expected to play a major role in the extra thermostabilization of proteins from organisms living under extreme temperature conditions (Ragone and Colonna, 1995). A recent study on protein-solvent interactions of several crystal structures of proteins from meso- and thermophiles indicated that structures of the latter are characterized by a high degree of optimization of the hydrophobic interactions or by highly optimized charge-charge interactions, in cases where the optimization of hydrophobic interactions was not sufficiently high (Spassov et al., 1995).

Temperature adaptation of biologically active proteins such as enzymes encompasses not only their stability characteristics, but also their functional properties. Enzymes are dynamic molecules and a certain degree of structural flexibility is essential for them to perform their function as biocatalysts (Vihinen, 1987, Shoichet et al., 1995). Thus optimizing a function of an enzyme under given temperature conditions requires a compromise between two often opposing factors, structural rigidity and flexibility of the protein molecule (Jaenicke, 1991). Enzymes performing the same catalytic function, but adapted to different habitat temperatures, appear to have about the same conformational flexibility at their respective temperature optima (Vihinen, 1987). For enzymes from thermophilic microorganisms, that appear to have very rigid, compact structures, with a tightly packed hydrophobic core and maximal exposure of hydrophilic residues at the surface, the conformational flexibility optimal for activity is reached only at elevated temperatures. For cold-adapted enzymes the optimal flexibility must be achieved at low temperatures, and it has been proposed that these enzymes must therefore possess flexible protein structures to compensate for the lower thermal energy provided by the low temperature habitat (Hochachka and Somero, 1984). As this would require weakening or alterations of some intramolecular interactions, the structural stability of cold-adapted enzymes is expected to be diminished in comparison to their counterparts adapted to higher temperatures. Understanding the factors that are responsible for the decreased thermal stability of these enzymes may therefore provide an important insight into the mechanisms underlying their cold-adaptation.

Proteins are denatured by cold as well as by heat (Privalov 1990), although the source of destabilization of native proteins causing them to unfold may be different at the two temperature extremes (Franks, 1995). Adaptation of enzymes to low temperatures may therefore require "stabilization" of their native conformations against cold-denaturation. Unfolding of proteins at low temperatures has been found to occur most readily with the most hydrophobic proteins (Creighton, 1991). It remains to be seen however, whether cold-adapted proteins rely less on hydrophobic interactions for stability than enzymes from meso- and thermophiles. The subject of stability of proteins at low temperatures has recently been reviewed and commented on by Franks (1995).

COLD-ADAPTED SERINE PROTEINASES

Studies on enzymes from coldwater fishes

General kinetic characteristics. Catalysis by enzymes must in general involve some movement of residues within the enzyme molecule as the substrate(s) goes through structural alterations along the reaction time coordinate. The ease with which such movement can occur may be one of the determinants of catalytic efficiency. An illustration of this is the observation that cooling to room temperature abolishes activity of many thermophilic enzymes, presumably because thermal energy in the surroundings is insufficient to provide energy for the relative movements of enzyme-groups, making the enzyme molecule too rigid. An example is the metalloproteinase thermolysin, for which a negative correlation between activity and stability was observed. A difference in flexibility around the active site was suggested to be the key to this correlation (Kidokoro, et al., 1995).

Organisms living in cold habitats have adapted their metabolic rates to the challenge of low environmental temperature by either producing more enzyme molecules or by increasing the catalytic potential of individual molecules. Thus, the catalytic efficiency of enzymes from such organism is higher than measured for similar enzymes from warm-blooded animals when placed in the cold. A number of studies have demonstrated that reactions catalyzed by enzymes originating in poikilothermic organism such as cold-water bacteria or fish display lower activation energies compared to similar enzymes from warm-blooded animals (Davail, et al., 1994; Hochachka and Somero, 1984).

Kinetic properties of digestive serine endopeptidases have been studied from a number of fish species living in cold or temperate waters. Generally, the major difference in catalytic properties of these enzymes as compared with enzymes from warm-blooded animals is manifested in higher k_{cat} values, whereas K_m values are more often nearly unchanged. The most appropriate kinetic constant to compare as a measure of catalytic efficiency, particularly when subsaturating substrate concentrations prevail, is k_{cat}/K_m. This apparent second order rate constant sets a lower limit on the rate constant for enzyme-substrate association and is a measure of the specificity for competing substrates and the efficiency of the catalytic apparatus.

Table 1 summarizes kinetic data obtained for trypsin from several cold-water fish species and compares these with the kinetic properties of bovine trypsin using the synthetic substrate benzoyl-arginine p-nitroanilide (BzArg-pNA). It is apparent from this comparison that the catalytic efficiency (k_{cat}/K_m) of trypsins from species living in cold environments is markedly higher than observed for trypsin from a warm-blooded species, namely cattle. The ratio of k_{cat}/K_m values determined for the various fish species divided by that of bovine trypsin ranges from about two-fold to twenty eight-fold. In the case of most of these trypsins a significant contribution to the increase in catalytic efficiency (k_{cat}/K_m) results from lowering of K_m values, a situation that is rarely observed with the other members of this family of serine digestive proteinases (chymotrypsin and elastase) . Typically, k_{cat} values of the cold-adapted trypsins increased two- to eight-fold as compared with bovine trypsin when the synthetic substrate was used. Similar differences in turnover rates have been observed with native protein substrates (Kristjánsson, 1991) indicating that this difference in catalytic ability is relevant to the natural function of these enzymes (Figure 1). The large difference in K_m values observed

Table 1. Kinetic parameters determined for trypsins from cold-adapted fish and krill using BzArg-NH-pN as substrate and comparison with data for bovine trypsin taken from each study. Anionic trypsins are here designated as those with pI values below 7.0 and cationic trypsin with pI >9.0.

Species	Type	K_m (mM)	k_{cat} (1/sec)	k_{cat}/K_m (mM/sec)	k_{cat}/K_m ratio fish/bovine	°C assay	References
Trout (Salmo gairdneri)	Anionic	0.20	0.85	4.2	6.0	25	[Genicot et al., 1988]
vs. Bovine	Cationic	2.44	1.8	0.7	1	25	
Greenland cod (Gadus ogac)	Anionic?	1.67	3.5	2.1	4.1	25	[Simpson et al., 1989ab]
Cunner (Tautugolabrus adspersus)	Anionic?	0.73	3.0	4.1	8.2	25	
vs. Bovine	Cationic	1.02	0.5	0.5	1	25	
Atlantic cod (Gadus morhua)	Anionic I	0.077	4.0	51.9	16.7	25	[Ásgeirsson et al., 1989]
	Anionic II	0.094	1.9	20.2	6.5	25	
	Anionic III	0.102	0.7	6.8	2.2	25	
vs. Bovine	Cationic	0.650	2.0	3.1	1	25	
Rainbow trout (Oncorhynchus mykiss)	Anionic	0.077	3.28	42.6	11.8	20	[Kristjánsson, 1991]
vs. Bovine	Cationic	0.455	1.68	3.6	1	20	
Mullet (Mugil cephalus)(a)	Anionic?	0.49	6.55	13.4	3.9	25	[Guizani et al., 1991]
vs. Bovine	Cationic	0.88	3.0	3.4	1	25	
Atlantic salmon (Salmon salar)	Anionic I	0.050	1.77	35.4	24.9	20	[Outzen et al., 1996]
	Cationic	0.30	0.80	2.67	1.9	20	
vs. Bovine	Cationic	0.50	0.71	1.42	1	20	
Antarctic krill (Euphausia superba)	Anionic I	0.03	0.07	2.3	3.5	25	[Osnes and Mohr, 1985]
	Anionic II	0.04	0.74	18.5	28.5	25	
	Anionic III	0.04	0.45	11.3	17.4	25	
vs. Bovine	Cationic	0.939	0.61	0.65	1	25	

(a) k_{cat} values given per minute by authors here calculated per second.

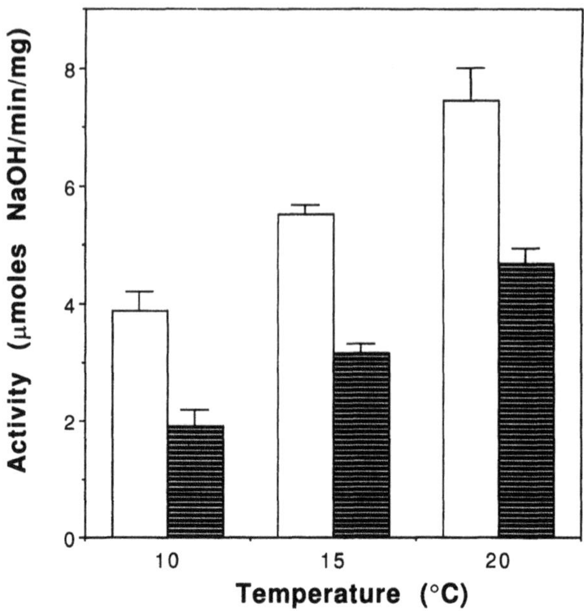

Figure 1. Comparison of caseinolytic activities of rainbow trout (open bars) and bovine trypsin (shaded bars). Activity is expressed as micromoles of OH- consumed per minute per milligram of enzyme at each temperature at pH 8.0, measured with a pH stat. (Adapted from Kristjánsson, 1991, with permission).

when using BzArg-pNA, a substrate which only occupies the P1 position with an amino acid residue, may indicate that the cold-adapted trypsins are capable of stronger binding of substrates in the primary substrate binding pocket compared with trypsins adapted to work in warm-blooded species.

It is noticeable that both anionic and cationic trypsins were found in Atlantic salmon, but only the anionic form displayed cold-adaptive behaviour, i. e. a higher k_{cat}/K_m value and lower stability in comparison to the mammalian enzymes (Outzen, et al., 1996). It was suggested that the higher catalytic efficiency observed for salmon trypsin might arise from the anionic character of the enzyme and the distribution of negatively charged residues in the structure (Outzen, et al., 1996). Although many of the fish serine proteinases, such as trypsins and chymotrypsins, are indeed anionic (Ásgeirsson and Bjarnason, 1991; Kristjánsson and Nielsen, 1992) the fish elastases have very alkaline pI values similar to the mammalian serine proteinases and yet display psychrophilic behavior (Ásgeirsson and Bjarnason, 1993; Bassompierre, et al., 1993; Berglund, et al., 1995; Gildberg and Øverbø, 1990).

Another point worth mentioning is that trypsin seems to be the only digestive serine proteinase where lowering of K_m is a significant part of the mechanism of cold-adaptation. In the case of Atlantic cod and rainbow trout, the kinetic properties of both chymotrypsins (Ásgeirsson and Bjarnason, 1991; Kristjánsson and Nielsen, 1992) and elastase (Ásgeirsson and Bjarnason, 1993; Bassompierre, et al., 1993; Gildberg and Øverbø , 1990) have also been studied. The K_m values determined for these enzymes with appropriate synthetic substrates were found to be very similar to those of comparative enzymes from warm-blooded animals. However, as with the trypsins, k_{cat} values were consistently 2-4 times larger, when protein substrates were employed (Bassompierre, et al., 1993; Gildberg and Øverbø, 1990; Kristjánsson and Nielsen, 1992).

Structural considerations. The active site of the digestive serine proteases (trypsin, chymotrypsin, elastase) is positioned between two six-stranded β-barrel domains (Figure 2). The fish serine proteinases are similar in this respect to the mammalian enzymes as exemplified by salmon trypsin (Smalås, et al., 1994). The fish enzymes maintain the same pattern of disulfide bridges which notably do not crosslink the two domains. There are six disulfide bonds in trypsin, five in chymotrypsin and four in elastase. The catalytic residues in trypsin are positioned in loops that face into the interdomain cavity. The N-terminal domain projects the His57 and Asp102 residues of the catalytic triad on one side into the active site cleft, whereas the catalytic Ser195 and the primary binding site for substrate P_1 side-chains are located on the opposite side as part of the C-terminal domain. The polypeptide chain traverses three times the interdomain cleft and in concert with hydrogen bonding and hydrophobic contacts between the domains, this holds the two domain structures together (Bode and Huber, 1986).

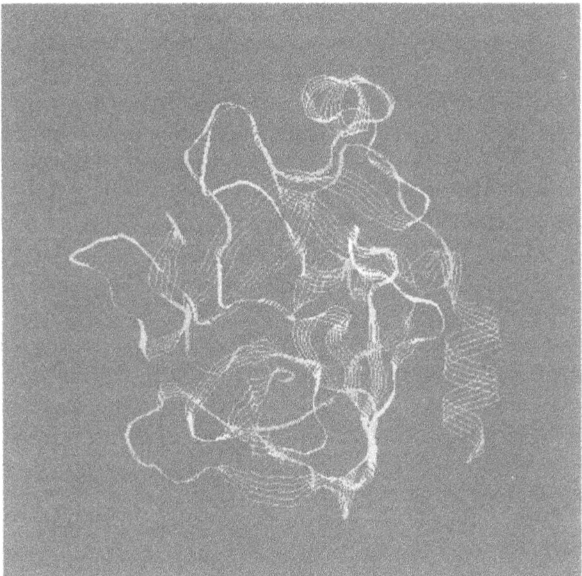

Figure 2. General structure of Atlantic salmon trypsin drawn with the RasMol computer program using coordinates from Smalås et al. (1994). The view is into the active site cleft and clearly shows the two-domain structure of the enzyme. The lower half is the N-terminal domain which carries the Asp102 and His57 of the catalytic triad, whereas the specificity pocket and Ser195 are in the upper C-terminal domain. The C-terminal helix is on the extreme right of the figure. The sparse distribution of strands connecting the two domains can also be appreciated.

The positioning of the catalytic site in a cleft between two quite rigid domains opens the possibility for varying the relative positioning of residues taking part in the catalytic process by domain movement. The key question, which still remains largely unanswered, however, is whether such movement plays a role in the catalytic mechanism of the serine proteinases. And if so, which kind of movements might be altered in order to achieve the higher catalytic rates observed in cold-adapted enzymes. It would follow to ask where in the structure such movement could take place and which residues are critical in allowing or facilitating such mobility of the protein structure. The formation of the acyl-enzyme intermediate in the serine proteinases requires significant torsional alterations in substrate structure which probably are facilitated through dynamic fluctuations of protein residues during binding (Wells, et al.,

1994). The importance of flexible surface loops in controlling kinetic properties of serine proteinases such as chymotrypsin have also been demonstrated (Hedstrom, et al., 1992).

If the positioning of groups around the peptide bond to be cleaved in the substrate could be forced out of the normal planar arrangement, the bond would attain a character of ester-bonds and be more easily hydrolyzed. A theory proposing that the relative movement of domains in the serine proteinases could achieve this has been proposed (Dufton, 1990). Peptide substrates would bind to one domain on the N-terminal side of the scissile bond and to the other domain with the C-terminal half. Thus, they might be poised for movement which might strain the substrate at the site of hydrolysis. However, no direct crystallographic data is available to support this kind of movement of domains. It should also be born in mind that the kinetic characteristics indicative of cold-adaptation can be detected using small ester- or amide derivatives of amino acids and peptides which do not make much contact with the enzyme, particularily on the part forming the first leaving group. Recent comparison of salmon trypsin with bovine trypsin by crystallography and molecular dynamics simulations indicate that there are only small differences in relative orientations of the two domains indicating little relative motion. However, the simulation time might have been too short to detect such motion (Heimstad, et al., 1995).

Loops are the most mobile parts of enzyme structures and also the parts where amino acid substitutions are most frequent. The variability in structure within the serine protease family is greatest in loops connecting the β-strands. The loops which are direct extensions of residues forming the specificity pocket were significantly more mobile in salmon trypsin as compared with bovine trypsin, and slightly higher fluctuation was observed in the salmon structure for the N-terminal loop, which is one of three strands connecting the two domains of trypsin (Heimstad, et al., 1995). Serine proteinases could make reactions go faster by forcing a better substrate alignment with the catalytic apparatus, for example by additional hydrogen bonding to residues adjacent to those directly involved in the tetrahedral intermediate. Studies using Raman spectroscopy have indicated that a very small movement of the carboxylic oxygen of substrates into the oxyanion hole in the active sites of serine proteinases can siginificantly affect catalytic rate (Tonge and Carey, 1992). Slight movement of catalytic groups could in a similar way create enhancement of reaction rates in the cold-active enzymes. Molecular dynamic simulation showed that the dynamic behaviour of some important residues of the catalytic site were different in the cold-active salmon trypsin compared with bovine trypsin (Heimstad, et al., 1995). Thus, the orientation of Asp102 seemed less stable in the latter enzyme where it went through rapid 180 °C "flips" of the carboxyl group. This movement of Asp102 was not observed in the salmon trypsin structure which might make the charge-relay pair of Asp102 and His57 act as a stronger base in catalysis by the salmon enzyme (Heimstad, et al., 1995).

In comparing the structures of salmon and bovine trypsins, Smalås and coworkers (1994) noted a difference in the distribution of charged residues on the surface of the two proteins, that might help to explain higher substrate binding affinity of the fish enzyme. The more anionic character of the salmon enzyme was found to be especially pronounced in the C-terminal domain on the same side as the entrance to the active site cleft . It was suggested that this negatively charged region of the molecule might help in locating or guiding the positively charged arginine and lysine side chains into the binding cleft. This could give rise to higher substrate binding affinity and hence the lower K_m that were obtained for the cold-adapted trypsin (Smalås et al., 1994).

Amino acid substitutions. Certain amino acid substitutions are known to affect the flexibility of the polypeptide chains and thus alter their stability. Thus introduction of prolines into mobile regions of proteins is expected to restrict available backbone configurations and thus increase rigidity, whereas glycines as a result of their high rotational freedom would increase chain flexibility. Matthews et al. (1987) have used this rational to engineer more stable variants (Gly 77 → Ala and Ala 82 → Pro) of T4 lysozyme. A recent example using a mesophilic thermolysin-like proteinase demonstrated a correlation between higher thermal stability and amino acid exchanges that included the removal of glycines and/or addition of prolines in strategic places (Eijsink, et al., 1995). A dramatic increase in thermal stability was obtained by very few mutations in one particular flexible loop region of the molecule. This flexible loop is the site where autolysis is initiated after partial thermal unfolding of this part of the molecule. The specific activities were reported similar for all the mutants in the study (Eijsink, et al., 1995). The serine proteinase, subtilisin BPN', has also been stabilized by reinforcing the structure of an extended surface loop. In that case, however, stabilization was achieved by increasing the rigidity of the autolysis loop by incorporating a calcium binding site into it (Braxton and Wells, 1992). Such solvent-exposed surface loops may thus be critical sites for determining the thermal stability of broad specificity proteinases.

When comparing the amino acid composition and sequences of cold-adapted serine proteinases from fish and their mammalian counterparts it is noteworthy that the former generally have a higher content of methionine residues (Gudmundsdottir et al., 1993,1994, 1996). Both Atlantic cod and salmon trypsin contain six methionines, at the same locations in their sequences (Gudmundsdottir et al., 1993, Smalås et al., 1994). Two of those residues, located in β-structures, are conserved in the bovine enzyme, but others are replaced by different amino acids. Methionine has a linear side chain with three freely rotating χ-angles, thus allowing greater flexibility of the polypeptide chain (Ponder and Richards, 1987). Two of the additional Met residues in the fish trypsins are located in relatively flexible parts; Met145 is located in the autolysis loop and Met242 is part of the C-terminal helix (Figure 2). These substitutions may add to the mobility of those regions of the cold-adapted enzymes. Another noteworthy difference in amino acid sequences of the fish and mammalian trypsins, that may have implications for flexibility of the proteins, are deletions of two proline residues in the fish enzymes. Pro28, which is part of the polypeptide chain that crosses the cleft between the two domains, is highly conserved in mammalian trypsins. This residue is substituted to Ala in both cod and salmon trypsins (Gudmundsdottir et al., 1993, Smalås et al., 1994). As proline residues fix dihedral angles between the α-carbons and the peptide nitrogens to a small range, they restrict main chain conformations and thus induce rigidity. The Pro28 → Ala exchange may therefore increase flexibility of this interdomain region of the fish enzymes. Another proline residue, Pro152, is lacking in the salmon and cod enzymes. This residue is located in the autolysis loop of the bovine trypsin, its deletion in the fish enzymes may thus increase flexibility in this loop, however with unknown consequences for either activity or stability of the enzymes.

Amino acid substitutions can affect surface charge of enzyme molecules and the effects of such changes should not be dismissed as irrelevant to events at the catalytic site. Surface charges have been found to influence the flexibility of chymotrypsin molecules as seen by applying electron paramagnetic resonance spectroscopy and molecular dynamics simulations (Affleck, et al., 1992; Hartsough and Merz, 1992). Motion in the vicinity of the spin-labelled Met192 and Ser195 residues decreased dramatically with decreasing dielectric constant of the solvent, indicating that changes in the electrostatic forces between charged residues of the protein could affect the dynamics of catalytic groups (Affleck, et al., 1992). Electrostatic forces may also affect the preferred orientation of amino acid side chains (Hartsough and Merz, 1992). It is well established that stereoselectivity of chymotrypsin changes in nearly dry organic solvents which might result from differences in mobility properties in the structure (Affleck, et al., 1992). Interestingly, long range effects from a surface charge to the active site in subtilisin were found to lower the activation energy of catalysis by 0.4 kcal/mol by stabilization of the catalytic transition state (Jackson and Fersht, 1993).

Experimental evidence shows that single point mutations can create enzymes with improved activity (Fersht and Wilkinson, 1985, Kubo et al., 1992, Shih and Kirsch, 1995). One might therefore ask if single amino acid substitutions could generally be expected to bring about the functionally significant alterations that are observed in cold-adapted enzymes. However, as the effects of individual single amino acid substitutions are usually found to be additive (Alber, 1989, Shih and Kirsch, 1995), it is likely that cold-adaptive changes may have come about gradually with cumulative contributions from several amino acid substitutions.

Enzyme activity has often been observed to decrease with increased stability and evidence suggests that enzymes are optimized for activity rather than stability (Stoichet et al., 1995). While there are several observations to support this view, a recent study of mutants of chicken lysozyme have pointed out that this may not be a general rule. A series of mutant amino acid residues (six in all) were constructed into chicken lysozyme which interestingly enhanced both thermal and chemical stability, as well as increasing specific activity of the enzyme by 2.5-fold (Shih and Kirsch, 1995). This somewhat unexpected change in catalytic function could be attributed to the cumulative higher activities of several of the constituent substitutions. However, a single amino acid substitution (Asp101 $\rightarrow$ Ser) increased the activity as much as eighty percent, as well as raising the melting temperature of the enzyme by 2.2 °C (Shih and Kirsch, 1995).

From the few selected examples discussed here it is clear that substitution of a single or few amino acid residues in an enzyme can profoundly affect both its activity and stability. Studies on cold-adapted enzymes using site-directed mutagenesis are still lacking but should offer new insight into the primary cause for their high catalytic activity.

Stability characteristics. Serine proteinases isolated from fish have generally been found to be more thermolabile and are also often found to be more acid labile than the corresponding enzymes from warm-blooded organisms (Ásgeirsson et al., 1989, Ásgeirsson and Bjarnason, 1993, Kristjánsson, 1991, Kristjánsson et al., 1995). In the case of Atlantic cod trypsins, the half-lives of anionic forms I and III were measured 29 and 19 minutes, respectively, at 50 °C and pH 7.3, but was less than 2 minutes for cod trypsin II. Under the same conditions the half-life of bovine trypsin was 51 minutes (Kristjánsson et al., 1993). Cod elastase loses half of its activity after heating for 10 minutes at 48 °C, whereas for the porcine enzyme the corresponding temperature was 63 °C (Figure 3) (Ásgeirsson and Bjarnason, 1993).

Comparisons of the crystal structure of an anionic form of salmon trypsin to that of bovine trypsin revealed some differences that may reflect the lower stability of the fish enzyme (Smalås et al., 1994). The two enzymes share 65% sequence homology and as mentioned previously their tertiary structures very similar. Furthermore, comparison of temperature factors (B-factors) did not indicate any overall differences in flexibility of the two structures (Smalås et al. 1994, Heimstad et al., 1995). Although differences in the average hydrophobicity between residues in the salmon and bovine enzyme were found to be relatively small, internal residues of the former tended to be less hydrophobic, but external residues more hydrophobic than in the bovine enzyme. Distinct differences were also noted in the interactions between the N-terminal and C-terminal domains in the two trypsins. Fewer hydrogen bonds exist between the domains of the fish enzyme and apparently interdomain hydrophobic interactions were also weaker, due to smaller hydrophobic residues in this region, leading to formation of cavities of the size of 3-4 methyl-groups, that may decrease the stability of the enzyme (Smalås et al., 1994). Differences were also observed in the structures in a contact region between a C-terminal helix and and the rest of the molecule (Figure 2). In the bovine enzyme there are several hydrogen-bonds formed between this helix and the N-terminal domain, interactions that are not present in salmon trypsin. Electron density maps of the salmon enzyme indicate that this part of the molecule may also be more flexible than in bovine trypsin. Weaker interactions in this region that forms like an "arm" along the "body" of the protein and thus makes a link between the C-terminal and the N-terminal domains of the molecule may therefore

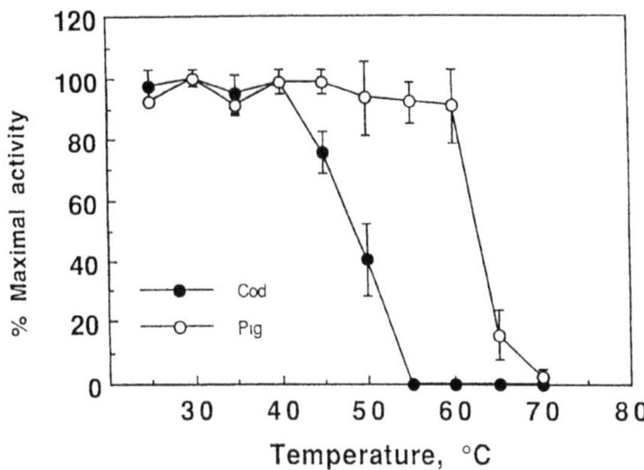

Figure 3. Effect of temperature on the stability of Atlantic cod elastase (●) and porcine elastase (o). Residual activity was measured after incubation in 0.2 M Tris and 10 mM CaCl2 for 10 min at the temperatures indicated (From Ásgeirsson and Bjarnason, 1993, with permission)

contribute significantly to lower thermal stability of the fish enzyme (Figure 2) (Smalås et al., 1994). Weakened electrostatic interactions involving the carboxylate group of the C- terminal residue and Lys-87 and Arg-107, were also suggested to contribute to the lower stability of the salmon enzyme.

Characterized fish serine proteinases are often found to be unstable under acidic conditions and are typically irreversibly inactivated at pHs below 5 (Ásgeirsson et al., 1989, Ásgeirsson and Bjarnason, 1993, Kristjánsson, 1991, Kristjánsson et al., 1995). Smalås and coworkers suggested that protonation of the C-terminal carboxylate group may contribute to the acid lability of salmon trypsin. As a result of fewer hydrogen bond contacts between the C-terminal helix and rest of the molecule, it can be envisaged that this ionic interaction involving the carboxylate plays a critical role in keeping the helix "arm" in place along the side of the molecule. Low pH might disrupt this interaction, resulting in an increased motility of the helix, with potentially dire effects for the integrity of the active enzyme structure (Smalås et al., 1994).

Calcium binding appears to stabilize fish serine proteinases similarly to other pancreatic serine proteinases (Kristjánsson, 1991, Kristjánsson and Niesen, 1992, Kristjánsson et al., 1995). These enzymes bind a calcium ion at a single site, located in a loop that appears to be a common feature in their structures (Bode and Schwager, 1975). The structure of the calcium binding site of the anionic salmon trypsin and bovine trypsin were found to be similar (Smalås, 1994). Calcium binding may however, stabilize the trypsins to a different extent toward autolysis, as the salmon trypsin did not show the same increase in autolysis in the absence of calcium as the bovine enzyme did (Smalås et al., 1994). Cod trypsins also differ from the bovine enzyme with respect to calcium stabilization. In the absence of calcium the rates of inactivation at 50 °C of the bovine enzyme increased about 17-fold, but that of cod trypsin isoforms I and III by 6 to 8 times. Calcium binding on the other hand had no apparent effect on the stability of cod trypsin II. Diminished calcium binding may therefore explain the thermolability of isoform II of cod trypsin (Kristjánsson and Bjarnason, unpublished results).

Studies on Psychrophilc bacteria

Relatively little information is available on properties of enzymes from psychrophilic bacteria. Recently, however, a few reports on enzymes from psychrophilic bacteria have appeared in the literature that had the objectives to elucidate mechanisms of their cold-adaptation (Davail et al., 1992, 1994, Feller et al., 1991, 1992, 1994, Rentier-Delrue, 1993), including one on a subtilisin-like serine proteinase secreted by the Antartic psychrophile *Bacillus* TA41 (called subtilisin S41) (Davail et al.,1994). This enzyme shares most of the enzymatic characteristics with the mesophilic subtilisins of other *Bacillus* species, but shows some cold-adaptive traits, such as a lower temperature optimum and lower thermal stability (Davail, et al., 1994). The amino acid sequence of the enzyme showed 52% residue identity with the mesophilic enzymes and residues in secondary structures, the active site region and the two calcium binding sites are well conserved. A model of the three-dimensional structure of the psychrophilic enzyme was built on the basis of this primary structure and the known structures for other subtilisins (Davail et al., 1994). Comparison of this model and the known subtilisin structures revealed several differences that might explain lower thermal stability and hence a more flexible structure of the cold-adapted enzyme (Table 2). Several ionic and aromatic interactions, present in the mesophilic and thermophilic subtilisins, are lacking in the cold-active enzyme. Ionic interactions are usually rather poorly conserved within evolutionary families, including subtilases (Siezen et al, 1991), suggesting that they do not play a critical role in the stability or the folding pathway of proteins (Barlow and Thornton,1983, Baldwin and Eisenberg, 1987). More numerous ionic interactions in thermitase, as compared to the mesophilic subtilisins, have though been implicated in enhanced thermal stability of that enzyme (Teplyakov et al., 1990).

Table 2. Comparison of main structural features expected to affect stability of psychrophilic subtilisin S41 in comparison to related meso- and thermophilic subtilases (adapted from Davail et al., 1994, with permission).

Parameter	S41	BPN'	Carlsberg	Savinase	Thermitase	Expected effect S41 stability
Amino acids	309	275	274	269	279	$-$ [a]
Asp content	21	10	9	5	13	(extended surface loops)
Ionic interactions	2	5	3	7	10	(hydrophilic surface)
Aromatic interactions	0	4	5	3	11	-
Kd for Ca^{2+}	10^{-6}	10^{-10}	10^{-10}	?	$<10^{-10}$	-
Disulfide bonds	Cys^{47}-Cys^{58}	0	0	0	0	0 (experimental)

[a] - denotes an expected decreases in stability.

There are two ionic interactions present in the model structure for the cold-adapted enzyme (Asp49-Lys94 and Glu197-Arg247). The Asp49-Lys94 interaction is not present the mesophilic subtilisins as residue 49 is a serine in those enzymes. Asp 49 is located at a start of a protruding loop close to the active site, that has been identified as one of the two main autolytic sites (Ala48-Ser49) in the subtilisins (Braxton et al., 1992). The Ser49 to Asp exchange in the cold-adapted enzyme may therefore stabilize it against autolysis. This salt-bridge is present in thermitase where it stabilizes a loop containing the extra calcium binding site presently known only in that enzyme (Siezen et al., 1991). The interaction between residues 197 and 247 is involved in stabilization of the weak calcium binding site and is also present in other subtilisins.

Calcium binding plays a critical role in stabilization of all known subtilases (Pantoliano et al., 1988, Braxton et al., 1992, Müller et al,1995, Genov et al., 1995) These enzymes bind calcium at two sites, one with high binding affinity, but the other with much lower affinity. Thermitase, additionally contains a third calcium binding site that contributes to increased thermal stability of the enzyme (Betzel et al.1990, Teplyakov et al., 1990). Calcium binding at the high affinity calcium site of the psychrophilic proteinase was significantly weaker than in the mesophilic enzymes and may therefore also contribute to lower stability of the enzyme (Table 2) (Davail et al., 1994).

Increased number of aromatic interactions has been given as one of the reasons for enhanced stability of thermitase (Teplyakov et al.,1990). As such interactions may add an extra 2.5-5.4 kJ mol^{-1} to the free energy of stabilization of a protein (Burley and Petsko, 1985, 1988), it is clear that the apparent absence of aromatic interactions in the psychrophilic subtilisin, may contribute significantly to decreased stability of the enzyme.

Subtilisin S41 also differed from its mesophilic counterparts in having an acidic isoelectric point (pI 5.3) due to higher Asp content and the molecular mass of the enzyme was also somewhat higher than for mesophilic subtilisins, as a result of seven sequence insertions. It was argued that the high Asp content, located especially in extended surface loops, may contribute to decreased stability of the cold-adapted enzyme, by providing a more hydrophilic surface that would give rise to improved solvent interactions and reduce the compactness of the molecule (Davail et al., 1994). An additional disulfide bond which is present in the psychrophilic, but not in meso-or thermophilic subtilisins does not appear to have any effect on its thermal stability (Davail et al., 1994).

Another example of a cold-adapted bacterial serine proteinase is that of an enzyme from a psychrotolerant *Vibrio* species. The enzyme, which is a subtilase belonging to class II, has a temperature optimum of about 10 °C lower than the mesophilic proteinase K from the fungus *Tritirachium album* Limber and 25 °C lower than that of a related proteinase from the thermophile *Thermus* strain IS-15 (Figure 4) (Kristjansson, unpublished results). The proteinase is also more thermolabile, by about 15 °C and 37 °C, as compared to proteinase K and a thermophilic proteinase from a *Thermus* strain IS-15, respectively. As with other subtilases, the cold-adapted proteinase is highly dependent on calcium binding for stability and disulfide bonds are essential for the integrity of the active protein structure. The *Vibrio*-proteinase contains four disulfide bonds whereas proteinase K has two and one free sulfhydryl group (Betzel et al., 1990, Müller et al., 1993). When comparing properties of the Vibrio-proteinase and proteinase K a noticeable difference was observed with regard to the reactivity of their disulfide bonds which may reflect structural differences related to temperature adaptation. When the proteinases were subjected to a reaction with excess sulfite in the presence of 2-nitro-5-thiosulfobenzoate (NTSB), but in the absence of a denaturant (Thannhauser et al., 1984), all the disulfides of the *Vibrio*-proteinase were cleaved and converted to thiosulfonates, leading to a loss in activity of the enzyme. With proteinase K, no cleavage of the disulfides occurred unless the reaction was carried out in the presence of the potent denaturant guanidinium thiocynate (Figure 5). As disulfide bonds in proteins are generally found to be solvent inaccessible (Srinivasan et al., 1990), this difference in reactivity of the disulfides of the two related proteinases, probably reflects difference in accessibility of these groups in the two proteins. The apparent difference in accessibility may reflect a difference in flexibility of the two protein structures. The psychrotrophic enzyme proposedly being more flexible, may allow the reagent more access to the disulfides in the protein, contrary to what may be envisioned for the more rigid protein structure of the mesophilic enzyme.

In addition to proteolytic enzymes, α-amylase, lipase and triose phosphate isomerase iso-

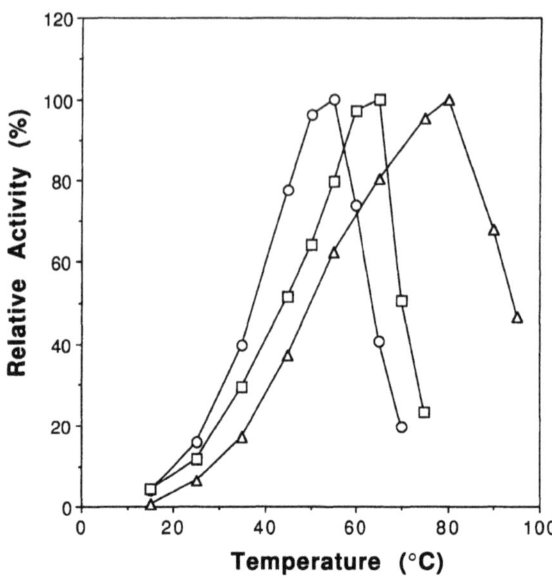

Figure 4. Effect of temperature on the proteolytic activity of a proteinase from the psychrotolerant *Vibrio* strain PA44 (o), proteinase K (□) and a proteinase from the thermophilic *Thermus* stain IS-15 (△). Activity was measured with 1% azocasein as a substrate in 100 mM glycine, 10 mM CaCl2, pH 9.5. Incubation was for 30 min.

lated from psychrophilic Antartic bacteria have also been studied with respect to cold-adaptation (Feller et al., 1991, 1992, 1994, Rentier-Delrue et al., 1993, Arpigny et al., 1993). The α-amylase from the psychrophile *Alteromonas haloplanctis* shows higher activity at lower temperatures and lower thermal stability, both cold-adaptive properties (Feller et al., 1992, 1994). A comparison of a molecular model for the psychrophilic α-amylase and the porcine enzyme, indicated a lack of several surface salt-bridges, fewer weakly polar interactions involving aromatic residues, as well as decreased hydrophobicity in the psychrophilic enzyme. The enzyme also had lower content of proline and arginine residues, as well as decreased binding affinities for Ca^{2+} and Cl^- ions as compared to porcine α-amylase (Feller et al., 1994). It was suggested that these changes in the structure in the psychrophilic enzyme could give rise to higher molecular flexibility required to compensate for the reduced reaction rates at low temperatures (Feller et al., 1992,1994). A similar strategy appears therefore to be involved in the psychrophilic adaptation of the α-amylase as observed in the studies on subtilisin S41. It has been suggested however, that the balance between the contribution of exothermically (electrostatic interactions that are reinforced at low temperatures) and endothermically formed interactions (hydrophobic interactions that are strengthened with temperature) may be more critical than the actual number of altered weak interaction in the adaptation to low temperatures (Feller et al, 1994, Davail et al., 1994).

CONCLUSIONS

Research on proteins from cold-adapted organisms is still at its early stages. Only few selected enzymes have been studied with the objective to elucidate the molecular mechanisms underlying cold-adaptation of proteins. In line with the original hypothesis it is likely that temperature adaptation involves adjustment of global or local flexibilities of the protein

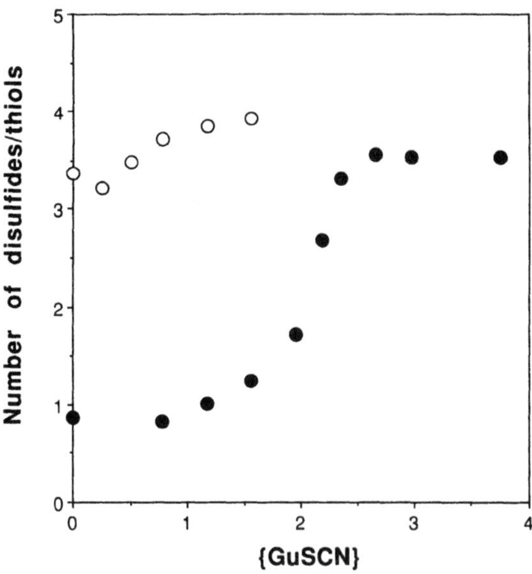

Figure 5. Effect of guanidine thiocyanate on the reactivities of disulfides in *Vibrio*-proteinase (o) and proteinase K (•) to thiosulfonation in the NTSB-reaction.

molecules to compensate for different thermal input in the various temperature habitats. Thus cold-adaptation would be expected to involve amino acid exchanges that would allow critical regions of the molecules to be more flexible, however without forfeiting too much of global stability of the protein structure. To understand temperature adaptation it is therefore required to pinpoint those critical amino acid exchanges. As seems to be the case for thermophilic proteins this may involve only minor alterations in several interactions at different locations in the proteins, although the possibility of few localized substitutions in critical regions of the protein structures, having prominent roles in adaptation, cannot be excluded. Strengthening of hydrophobic interactions and increased packing densities apparently play important roles in increased stability of thermophilic proteins. The question arises whether the opposite may be the case for psychrophilic proteins, i.e. does cold-adaptation involve some changes in packing densities in the proteins and does the hydrophobic interaction play less important role in maintaining stable structures of those proteins at low temperatures. In view of the temperature dependence of the hydrophobic effect, it is not unreasonable to assume that cold-adapted proteins may have to rely more on other types of interactions for stability to compensate for decreased strength of hydrophobic interactions. To answer these and other questions concerning cold-adaptation calls for more studies on structural and functional properties of proteins from psychrophilic organisms. As more sequence and higher level structural information becomes available, more rigorous interpretation of experimental observations can be done and would help directing a rational design of site-directed mutagenesis experiments that would aid in testing hypothesis concerning structural and functional aspects of cold-adaptation of proteins.

REFERENCES

Adams, M.W.W., Perler, F.B., and Kelly, R.M., 1995, Extremozymes: expanding the limits of biocatalysis. *Biotechnol.* 13:662-668.

Affleck, R., Haynes, C., and Clark, D., 1992, Solvent dielectric effects on protein dynamics. *Proc. Natl. Acad. Sci. USA* 89:5167-5170.

Alber, T., 1989, Mutational effects on protein stability. *Ann. Rev. Biochem.* 58:765-798.

Argos, P., Rossmann, M.G., Grau, U.M., Zuber, H., Frank, G., and Tratschin, J.D., 1979, Thermal stability and protein strucure. *Biochemistry* 18:5698-5709.

Arpigny, J.L., Feller, G., and Gerday, C., 1993, Cloning, sequence and structural features of a lipase from the antarctic facultative psychrophile *Psychrobacter immobilis* B10. *Biochim. Biophys. Acta* 1171:331-333.

Ásgeirsson, B., Fox, J.W., and Bjarnason, J.B., 1989, Purification and characterization of trypsin from the poikilotherm *Gadus morhua. Eur. J. Biochem.* 180:85-94.

Ásgeirsson, B., and Bjarnason, J.B., 1991, Structural and kinetic properties of chymotrypsin from Atlantic cod (*Gadus morhua*). Comparison with bovine chymotrypsin. *Comp. Biochem. Physiol.* 99B:327-335.

Ásgeirsson, B. and Bjarnason, J.B., 1993, Properties of elastase from Atlantic cod, a cold-adapted proteinase. *Biochim. Biophys. Acta* 1164:91-100.

Baldwin, R.L., and Eisenberg, D., 1987, Protein stability. In: Protein Engineering, pp. 127-148, Alan R. Liss, Inc., New York.

Barlow, D.J., and Thornton, J.M., 1983, Ion-pairs in proteins. *J. Mol. Biol.* 168:867-885.

Bassompierre, M., Nielsen, H.H., and Børresen, T., 1993, Purification and characterization of elastase from the pyloric caeca of rainbow trout (*Oncorhynchus mykiss*). *Comp. Biochem. Physiol.* 106B:331-336.

Berglund, G.I., Smalås, A., and Willasen, N.P., 1995, Purification and characterization of pancreatic elastase from Atlantic salmon (*Salmo salar*). Doctoral thesis. Department of Chemistry. University of Tromso, Norway.

Betzel, C., Teplyakov, A.V., Harutyunyan, E.H., Saenger, W., and Wilson, K.S., 1990, Thermitase and proteinase K: a comparison of the refined three-dimensional structures of the native enzymes. *Protein Eng.* 3:161-172.

Betzel, C., Klupsch, S., Papendorf, G., Hastrup, S., Branner, S., and Wilson, K., 1992, Crystal structure of the alkaline proteinase savinase^{TM} from *Bacillus lentus* at 1.4 Å resolution. *J. Mol. Biol.* 223:427-445.

Blake, P.R., Park, J.B., Zhou, Z.H., Hare, D.R., Adams, M.W.W., and Summers, M.F., 1992, Solution-state structure by NMR of zinc-substituted rubredoxin from the marine hyperthermophilic archaebacterium *Pyrococcus furiosus. Protein Sci.* 1:1508-1521.

Bode, W., and Schwager, P., 1975, The single calcium-binding site of crystalline bovine β-trypsin. *J. Mol. Biol.* 98:693-717.

Bode, W., and Huber, R., 1986, Crystal structures of pancreatic serine endopeptidases. Molecular and cellular basis of digestion. In: Molecular and Cellular Basis of Digestion, Desnuelle, P., Sjöström, H. and Norén, O., eds, pp.213-234.

Bode, W., Papamokos, E., and Musil, D., 1987, The high resolution X-ray crystal structure of the complex formed between subtilisin Carlsberg and eglin C, an elastase inhibitor from the leech *Hirudo medicinalis. Eur. J. Biochem.* 166:673-692.

Bond, J., and Butler, P.E., 1987, Intracellular proteases. *Ann. Rev. Biochem.* 56:333-364.

Bott, R., Ultsch, M., Kossiakoff, A., Graycar, T., Katz, B., and Power, S., 1988, The three-dimensional structure of *Bacillus amyloliquefaciens* subtilisin at 1.8 Å and an analysis of the structural consequences of peroxide inactivation. *J. Biol. Chem.* 263:7895-7906.

Braxton, S. and Wells, J.A., 1992, Incorporation of a stabilizing Ca^{2+}-binding loop into subtilisin BPN'. *Biochemistry* 31:7796-7801.

Burley, S. K. and Petsko, G. A. 1985, Aromatic-aromatic interactions: A mechanism of protein structure stabilization. *Science* 229:23-28.

Burley, S.K., and Petsko, G.A., 1988, Weakly polar interactions in proteins. *Adv. Protein Chem.* 39:125-189.

Böhm, G., and Jaenicke, R., 1994, Relevance of sequence statistics for the properties of extremophilic proteins. *Int. J. Peptide Protein Res.* 43:97-106.

Carter, P., and Wells, J.A., 1990, Functional interaction among catalytic residues in subtilisin BPN'. *Proteins* 7:335-342.

Chan, M.K., Mukund, S., Kletzin, A., Adams, M.W.W., and Rees, D.C., 1995, Structure of a hyperthermophilic tungstopterin enzyme, aldehyde ferredoxin oxidoreductase. *Science* 267:1463-1469.

Creighton, T.E., 1991, Stability of folded conformations. *Curr. Opin. Struct. Biol.* 1:5-16.

Davail, S., Feller, G., Narinx, E., and Gerday, C., 1992, Sequence of the subtilisin-encoding gene from an antarctic psychrotroph *Bacillus* TA41. *Gene* 119:143-144.

Davail, S., Feller, G., Narinx, E., and Gerday, C., 1994, Cold adaptation of proteins. Purification, characterization, and sequence of the heat-labile subtilisin from the antarctic psychrophile *Bacillus* TA41. *J. Biol. Chem.* 269:17448-17453.

Day, M., Hsu, B.T., Joshua-Tor, L., Park, J.B., Zhou, Z.H., Adams, M.W.W., and Rees, D.C., 1992, X-ray crystal structures of the oxidized and reduced forms of the rubredoxin from the marine hyperthermopphilic archaebacterium *Pyrococcus furiosus. Protein Sci.* 1:1494-1507.

Dill, K.A., 1990, Dominant forces in protein folding. *Biochemistry* 29:7133-7155.

Doig, A.J., and Williams, D.H., 1991, Is the hydrophobic effect stabilizing or destabilizing in proteins? The contribution of disulfide bonds to protein stability. *J. Mol. Biol.* 217:389-398.

Dufton, M.J., 1990, Could domain movements be involved in the mechanism of trypsin-like serine proteases? *FEBS Lett.* 271:9-13.

Eijsink, V.G.H., Veltman, O.R., Aukema, W., Vriend, G., and Venema, G., 1995, Structural determinants of the stability of thermolysin-like proteinases. *Nature Struct. Biol.* 2:374-379.

Eriksson, A.E., Baase, W.A., and Matthews, B.W., 1993, Similar hydrophobic replacement of Leu99 and Phe153 within the core of T4 lysozyme have different structural and thermodynamic consequences. *J. Mol. Biol.* 229:747-769.

Estell, D.A., Graycar, T.P., and Wells, J.A., 1985, Engineering an enzyme by site-directed mutagenesis to be resistant to chemical oxidation. *J. Biol. Chem.* 260:6518-6521.

Feller, G., Thiry, M., and Gerday, C., 1991, Nucleotide sequence of the lipase gene *lip2* from the antarctic psychrotroph *Moraxella* TA144 and site-specific mutagenesis of the conserved serine and histidine residues. *DNA Cell Biol.* 10:381-388.

Feller, G., Lonhienne, T., Deroanne, C., Libioulle, C., van Beeumen, J., and Gerday, C., 1992, Purification, characterization, and nucleotide sequence of the thermolabile α-amylase from the antarctic psychrotroph *Alteromonas haloplanctis* A23. *J. Biol. Chem.* 267:5217-5221.

Feller, G., Payan, F., Theys, F., Qian, M., Haser, R., and Gerday, C., 1994, Stability and structural analysis of α-amylase from the antarctic psychrophile *Alteromonas haloplanctis* A23. *Eur. J. Biochem.* 222:441-447.

Fersht, A.R. and Wilkinson, A.J., 1985, Fine structure-activity analysis of mutations at position 51 of tyrosyl-tRNA synthetase. *Biochemistry* 24:5858-5861.

Franks, F., 1985, Biophysics and Biochemistry at Low Temperatures. Cambridge University Press, Cambridge.

Franks, F., 1995, Protein destabilization at low temperatures. *Adv. Protein Chem.* 46:105-139.

Genicot, S., Feller, G., and Gerday, C., 1988, Trypsin from antarctic fish (*Paranatothenia magellanica Forster*) as compared with trout (*Salmo Gairdneri*) trypsin. *Comp. Biochem. Biophysiol.* 90B:601-609.

Genov, N., Filippi, B., Dolashka, P., Wilson, K. S., and Betzel, C., 1995, Stability of subtilisins and related proteinases (subtilases) *Int. J. Peptide Protein Res.* 45:391-400.

Gildberg A., and Øverbø, K., 1990, Purification and characterization of pancreatic elastase from Atlantic cod (*Gadus morhua*). *Comp. Biol. Physiol.* 97B:775-782.

Gudmundsdóttir, Á., Gudmundsdóttir, E., Óskarsson, S., Bjarnason, J.B., Eakin, A., and Craik, C., 1993, Isolation and characterization of cDNAs from Atlantic cod encoding two different forms of trypsinogen. *Eur. J. Biochem.* 217:1091-1097.

Gudmundsdóttir, Á., Óskarsson, S., Eakin, A.E., Craik, C.S., and Bjarnason, J.B., 1994, Atlantic cod cDNA encoding a psychrophilic chymotrypsinogen. *Biochim. Biophys. Acta.* 1219:211-214.

Gudmundsdóttir, E., Spilliaert, R., Yang, Craik, C.S., Bjarnason, J.B., and Gudmundsdóttir, Á., 1996, Isolation and characterization of two cDNAs from Atlantic cod encoding two distinct psychrophilic elastases. *Comp. Biochem. Physiol.* (in press)

Guizani, N., Rolle, R.S., Marshall, M.R., and Wei, C.I., 1991, Isolation, purification and characterization of a trypsin from the pyloric caeca of mullet *(Mugil cephalus)*. *Comp. Biochem. Physiol.* 98B:517-521.

Hartsough D.S., and Merz K.M., 1992, Protein flexibility in aqueous and nonaqueous solutions. *J. Am. Chem. Soc.* 114:10113-10116.

Hedstrom L., Szilagyi L., and Rutter W. J. 1992, Converting trypsin to chymotrypsin: The role of surface loops. *Science* 255:1249-1253.

Heimstad E.S., Hansen L.K., and Smalås A.O., 1995, Comparative molecular dynamics simulation studies of salmon and bovine trypsins in aqueous solution. *Protein Eng.* 8:379-388.

Hochacha, P.W., and Somero, G.N., (1984) Biochemical Adaptation, pp.377-422. Princeton University Press, Princeton, New Jersey.

Jackson, S.E., and Fersht, A.R., 1993, Contribution of long-range electrostatic interactions to the stabilization of the catalytic transition state of the serine protease subtilisin BPN'. *Biochemistry* 32:13909-13916.

Jaenicke, R., 1991, Protein stability and molecular adaptation to extreme conditions. *Eur. J. Biochem.* 202:715-728.

Jaenicke, R., and Zavodsky, P., 1990, Proteins under extreme physical conditions. *FEBS Lett.* 268:344-349.

Kidokoro, S., Miki, Y., Endo, K., Wada, A., Nagao, H., Miyake, T., Aoyama, A., Yoneya, T., Kai, K., and Ooe, S., 1995, Remarkable activity enhancement of thermolysin mutants. *FEBS Lett.* 367:73-76.

Korndörfer, I., Steipe, B., Huber, R., Tomschy, A., and Jaenicke, R., 1995, The crystal structure of holo-glyceraldehyde-3-phosphate dehydrogenase from the hyperthermophilic bacterium *Thermotoga maritima* at 2.5 Å resolution. *J. Mol. Biol.* 246:511-521.

Kristjánsson, M.M., 1991, Purification and characterization of trypsin from the pyloric caeca of rainbow trout (*Oncorhynchus mykiss*) *J. Agric. Food Chem.* 39:1738-1742.

Kristjánsson, M.M., and Nielsen, H.H., 1992, Purification and characterization of two chymotrypsin-like proteases from the pyloric caeca of rainbow trout. *Comp. Biochem. Physiol.* 101B:247-253.

Kristjánsson, M.M., Ásgeirsson, B., Jensson, H., Fox, J.W., Chlebowski, J.F., Gudmundsdóttir, E., Craik, C., Gudmundsdóttir, Á., and Bjarnason, J.B., 1992, Characteristics and protein engineering of cod trypsin, a psychrophilic marine proteinase. Abstract from the *3rd Nordic Conference on Protein Engineering,* August 12th-16th, 1992, Korpilampi, Finland.

Kristjánsson, M.M., Gudmundsdóttir, S., Fox, J.W., and Bjarnason, J.B., 1995, Characterization of a collagenolytic serine proteinase from the Atlantic cod (*Gadus morhua*). *Comp. Biochem. Physiol.* 110B:707-717.

Kubo, M., Mitsuda, Y., Takagi, M., and Imanaka, T. 1992, Alteration of specific activity and stability of thermostable neutral protease by site-directed mutagenesis. *Appl. Environ. Microbiol.* 58:3779-3783.

Matthews, B.W., Nicholson, H., and Becktel, W.J., 1987, Enhanced protein thermostability from site-directed mutations that decrease the entropy of unfolding. *Proc. Natl. Acad. Sci. USA* 84:6663-6677.

McPhalen, C.A., and James, M.N.G., 1988, Structural comparison of two serine proteinase-protein inhibitor complexes: eglin-C-subtilisin Carlsberg and CI-2-subtilisin Novo. *Biochemistry* 27:6582-6598.

Mendez-Arias, L., and Argos, P., 1991, Engineering protein thermal stability. Sequence statistics point to residue substitutions in α-helices. *J. Mol. Biol.* 206:397-406.

Murphy, K.P., Privalov, P.L., and Gill, S.J., 1990, Common features of protein unfolding and dissolution of hydrophobic compounds. *Science* 247:559-561.

Müller, A., Hinrichs, W., Wolfs, W. M., and Sanger, W. 1994, Crystal structure of calcium-free proteinase K at 1.5 Å resolution. *J. Biol. Chem.* 269:23108-23111.

Neurath, H., 1984, Evolution of proteolytic enzymes. *Science* 224:350-357.

Osnes, K.K., and Mohr, V., 1985, On the purification and characterization of three anionic, serine-type peptide hydrolases from antarctic krill, *Euphasia superba. Comp. Biochem. Physiol.* 82B:607-619.

Outzen, H., Berglund, G.I., Smalås, A., and Willassen, N.P., 1996, Cold-adaptation features of trypsins from Atlantic salmon (*Salmo salar*). *Comp. Biochem. Physiol.* submitted.

Pantoliano, M.W., Whitlow, M., Wood, J.F., Rollence, M.L., Finzel, B.C., Gilliland, G.L., Poulos, T.L., and Bryan, P.N., 1988, The engineering of binding affinity at metal ion binding sites for the stabilization of proteins: subtilisin as a test case. *Biochemistry* 27:8311-8317.

Pedersen, J.T., Olsen, O.H., Betzel, C., Eschenburg, S., Branner, S., and Hastrup, S., 1994, Cavity mutants of savinase[TM]. Crystal structures and differential scanning calorimetry experiments give hints of the function of the buried water molecules in subtilisins. *J. Mol. Biol.* 242:193-202.

Ponders, J.W., and Richards, F.M., 1987, Internal packing and protein structural classes. *Cold Spring Harbor Symposia on Quantitative Biology*, Vol LII., 421-428. Cold Spring Harbor, New York.

Privalov, P.L., 1990, Cold denaturation of proteins. *Crit. Rev. Biochem. Mol. Biol.* 25:281-305.

Ragone, R., and Colonna, G., 1995, Do globular proteins require some structural peculiarity to best function at high temperatures? *J. Am. Chem. Soc.* 117:16-20.

Rentier-Delrue, F., Mánde, S.C., Moyens, S., Terpstra, P., Mainfroid, V., Goraj, K., Lion, M., Hol, W.G.J., and Martial J.A., 1993, Cloning and overexpression of the triosephosphate isomerasee genes from psychrophilic and thermophilic bacteria. Structural comparison of the predicted protein sequences. *J. Mol. Biol.* 229:85-93.

Rheinecker, M., Eder, J., Pandey, P.S., and Fersht, A.R., 1994, Variants of subtilisin BPN' with altered specificity profiles. *Biochemistry* 33:221-225.

Shih, P., and Kirsch, J.F., 1995, Design and structural analysis of an engineered thermostable chicken lysozyme. *Protein Sci.* 4:2063-2072.

Shoichet, B., Baase, W. A., Kuroki, R., and Matthews, B.W., 1995, A relationship between protein stability and protein function. *Proc. Natl. Acad. Sci. USA* 92:452-456.

Siezen, R.J., de Vos, W.M., Leunissen, J.A.M., and Dijkstra, W., 1991, Homology modelling and protein engineering strategy of subtilases, the family of subtilisin-like serine proteinases. *Protein Eng.* 4:719-737.

Simpson, B.K., Simpson, M.V., and Haard, N.F., 1989, On the mechanism of enzyme action: Digestive proteases from selected marine organisms. *Biotechnol. Appl. Biochem.* 11:226-234.

Simpson, B.K., Smith, J.P., Yaylayan, V., and Haard, N.F., 1989, Kinetic and thermodynamic characteristics of a digestive protease from Atlantic cod. *J. Food Biochem.* 13:201-213.

Smalås, A.E., Hjemstad, E.S., Hordvik, A., Willassen, N.P., and Male, R., 1994, Cold adaptation of enzymes: structural comparison between salmon and bovine trypsins. *Proteins* 20:149-166.

Spassov, V.Z., Karshikoff, A.J. and Ladenstein, R., 1995, The optimization of protein solvent interactions: thermostability and the role of hydrophobic and electrostatic interactions. *Protein Sci.* 4:1516-1527.

Strausberg, S., Alexander, P., Wang, L., Schwarz, F., and Bryan, P., 1993, Catalysis of a protein folding reaction: thermodynamic and kinetic analysis of subtilisin BPN' interactions with its propeptide fragment. *Biochemistry* 32:8112-8119.

Srinivasan, N., Sowdhamini, R., Ramakrishnan, C., and Balaram, P., 1990, Conformation of disulfide bridges in proteins. *Int. J. Peptide Protein Res.* 36:147-155.

Thannhauser, T.W., Konishi, Y., and Scheraga, H.A., 1984, Sensitive quantitative analysis of disulfide bonds in polypeptides and proteins. *Anal. Biochem.* 138:181-188.

Teplyakov, A.V., Kuranova, I.P., Harutyunyan, E.H., Vainshtein, B.K., Frömmel, C., Höhne, W. E., and Wilson, K.S., 1990, Crystal structure of thermitase at 1.4 Å. *J. Mol. Biol.* 214:261-279.

Tonge, P.J., and Carey, P.R., 1992, Forces, bond lengths, and reactivity: Fundamental insight into the mechanism of enzyme catalysis. *Biochemistry* 31:9122-9125.

Vihinen, M., 1987, Relationship of protein flexibility to thermostability. *Protein Eng.* 1:477-480.

Wells, J.A., and Estell, D.A., 1988, Subtilisin- an enzyme designed to be engineered. *Trends Biochem. Sci.* 13:291-297.

Wells, G.B., Mustafi, D., and Makinen, M.W., 1994, Structure at the active site of an acylenzyme of alpha-chymotrypsin and implications for the catalytic mechanism - an electron nuclear double resonance study. *J. Biol. Chem.* 269:4577-4586.

Zuber, H., 1988, Temperature adaptation of lactate dehydrogenase. Structural, functional and genetic aspects. *Biophys. Chem.* 29:171-179.

ENZYMATIC MODIFICATION OF FOOD PROTEINS
TO IMPROVE THE FUNCTIONAL PROPERTIES

Yoshiro Kamata

Miyagi University of Education
Sendai 980, Japan

INTRODUCTION

Several problems exists today in world food supply and consumption. For example, over intake of calorie, especially fat calories, in advanced countries is a serious problem. There is a high demand for food products that are high in quality and less in saturated fat and cholesterol. In the western countries, the fat problem may be overcome by lowering the intake of meat products. In the Asian countries, however, the problem is not overconsumption of meat products but the lack of it. The demand for meat products nonetheless is increasing in these countries because of rapid improvements in the standard of living. One of the solutions to alleviate this situation is the introduction of vegetable or milk protein-based simulated meat products. Proteins are the most important components in fabricated foods. Several proteins, including milk, egg and soy proteins, are already used widely in a variety of foods. However, more information on the physicochemical properties of these proteins are needed to improve their functionality and to extend their utilization in several fabricated food products. Chemical and enzymatic modification of food proteins will be an excellent approach to improve their functionality in a wide variety of foods. We have examined the efficacy some naturally occurring enzymatic reactions, including partial hydrolysis by proteases, crosslinking by transglutaminase, and a browning reaction for improving the functional properties of proteins. We have also studied the use of immobilized enzymes for modifying the functionalities of food proteins.

Food materials modified by either enzymatic or chemical methods must be safe for human consumption. Therefore, the methods or the reactions for food protein modification must be selected very carefully. If immobilized enzyme technology is to be used, then one should consider using edible support materials or systems for enzyme immobilization. Although the support materials leaching into a product, especially liquid-type products, can be easily removed by filtration, it is likely that some essential nutrient components that might be present as particulated material in the products also might be inadvertently removed by filtration. If edible support materials are used in such a reaction, we may be able to eliminate the filters, detectors or similar systems for preventing the contamination of food products.

LIMITED PROTEOLYSIS OF SOYBEAN GLOBULINS

We studied the use of proteolytic enzymes for improving the functional properties of soybean proteins. Of course several proteolytic enzymes have been used on several proteins for this purpose in food industry; however, there are very few reports on the application of limited proteolysis. The use of chymosin in cheese production is a well known example of limited proteolysis. Moreover, we used the limited proteolysis as a probe to understand the conformation of soybean globulins.

We studied the effects of limited proteolysis on structure and functionality of soy proteins, i.e., Glycinin and β-conglycinin, especially glycinin. Glycinin is a large molecule with a molecular weight of about 290,000~320,000 dalton (Koshiyama and Fukushima, 1976). It is composed of six subunits (Kitamura et al., 1976; Staswicket al., 1984; Marco et al., 1984) and has a rigid globular conformation. A detailed knowledge of the effects of glycinin conformation on its digestion by proteases is of much interest. It seems desirable to study how the native structural features of this protein affect the availability of its peptide bonds to cleavage by enzymes of known specificity.

Several authors have studied the action of proteolytic enzymes on native proteins. In the action of trypsin on myosin, Mihalyi and Harrington (1959) postulated the existence of two parallel first-order reactions with very different rate constants. They suggested that the enzyme attacks the random-coil regions of the molecule at a much higher rate than the "folded areas." Cowgill (1975) exposed myosin to attack by five proteolytic enzymes and in each case obtained a single high molecular weight fragment. Formation of the fragment was related to the stable of its conformation.

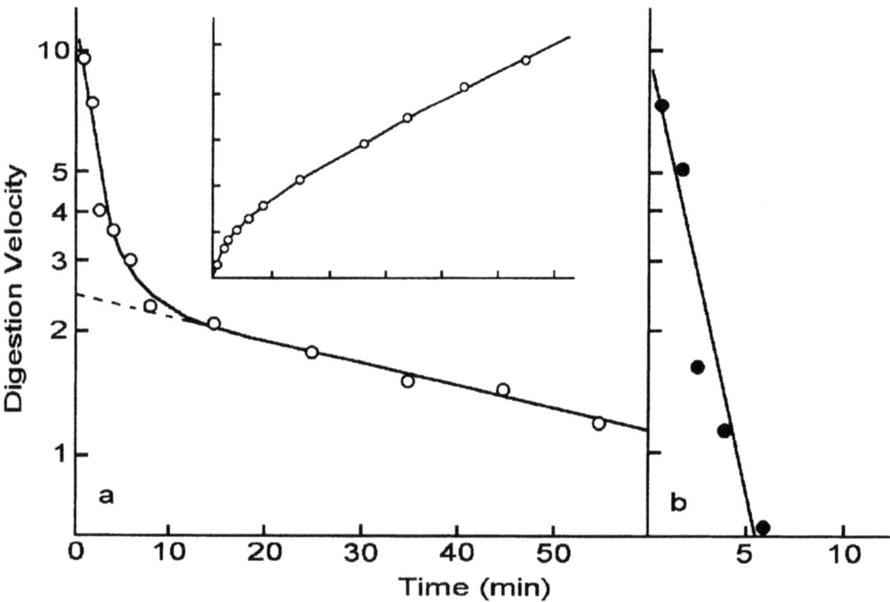

Figure 1. Graphical analysis of a typical pH-stat curve. Insert is a tryptic action on native glycinin followed by the pH-stat (adapted from Kamata and Shibasaki, 1978b with permission.)

We studied the action of trypsin on soy glycinin (Kamata and Shibasaki, 1978a; Kamata and Shibasaki, 1978b; Kamata et al., 1979a; Kamata et al., 1979b; Kamata et al., 1980; Kamata et al., 1982; Kamata et al., 1991a). The course of the reaction was followed over a long time period and the rate and extent of proteolysis were followed by a pH-stat (Fig. 1, insert; Kamata and Shibasaki,1978b). Initially, digestion proceeded very rapidly. Figure 1 shows a typical graphical analysis of glycinin digestion with trypsin. Representation of log velocity as a function of time gives a straight line (later section of the plot, a). By plotting the difference curve against time gives another straight line (b). The slope of each line represents the corresponding velocity constant of each reaction. The results of the kinetic analysis show that two independent reactions with different rate constants take place simultaneously. Stable digestion intermediates were expected from this reaction type. Disc gel electrophoresis shows a few intermediate bands and one of these intermediates (glycinin-T) is predominant under a high ionic strength condition (Fig. 2; Kamata and Shibasaki, 1978b). The electrophoretic mobility of glycinin-T is a little larger than that of glycinin. This is attributed to a more rigid conformation of glycinin under the condition. The reaction was stopped with soybean trypsin inhibitor after glycinin was completely converted to glycinin-T; the glycinin-T can be easily isolated by removing the enzyme and the inhibitor using a gel filtration column.

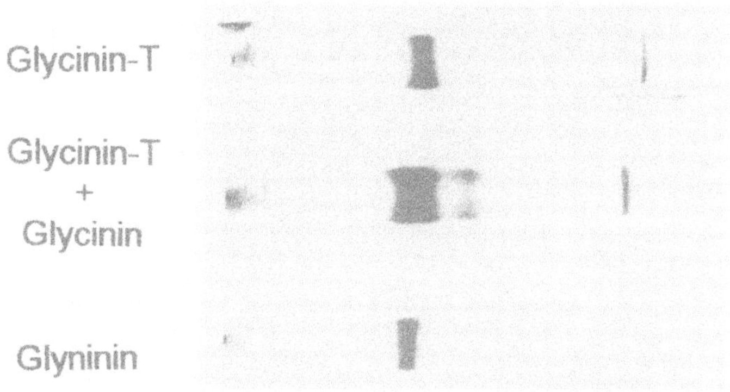

Figure 2. Disc gel electrophoresis of glycinin-T (adapted from Kamata and Shibasaki, 1978b with permission.)

Glycinin consists of two protomers; each protomer contains three subunits, which in turn consist of an acidic polypeptide chain and a basic polypeptide chain linked via disulfide bridges. Each acidic polypeptide chain is linked with the specific partner of the basic polypeptide chain through disulfide bridges, because these acidic-basic polypeptide chains pairs are generated from a precursor chain by posttranslational modification (Kitamura et al., 1976; Staswicket al., 1984; Marco et al., 1984).

Glycinin-T consists of six fragment groups originating from the subunits (Kamata and Shibasaki, 1978a; Kamata and Shibasaki, 1978b; Kamata et al., 1991a). Each group consists of a basic polypeptide chain, a T fragment that is connected to the basic polypeptide chain with a disulfide bond and a P fragment. When the subunit fragments are analyzed by SDS-urea polyacrylamide gel electrophoresis, some well-defined fragments other than the glycinin-T's fragments could be seen during the course of the fragmentation to glycinin-T (Kamata and Shibasaki, 1978a). The fragmentation process from glycinin to glycinin-T seemed to be very complex, because glycinin has a very complex oligomeric structure.

The kinetics of the fragmentation were followed by SDS-PAGE electrophoresis and densitometric gel scan. We classified these glycinin subunits into three groups (IS-1 ~ IS-3). It appears that fragmentation of glycinin subunits follow two main pathways: one is IS-3 (A3B4) fragmentation and the other is IS-1 (A1aB2) and IS-2 (A2B1a and A1bB1b) fragmentation. The isolation of the fragments of glycinin-T followed by comparison of the basic polypeptide chain pairing in them reveals the origin of the tryptic fragments of the subunit (IST).

Figure 3 (Kamata et al., 1991a) shows the time course for the formation of the fragments of glycinin. Glycinin is composed of three different subunit groups (IS-1, IS-2, and IS-3) which consist of an acidic polypeptide chain and a basic polypeptide chain. Only the acidic polypeptide chains are degraded and the fragmentation processes of IS's are similar. Initially, IS's are degraded into IST-1 or IST-2 followed by degradation to IST-3 or IST-4. The latter step is related to P fragments generation.

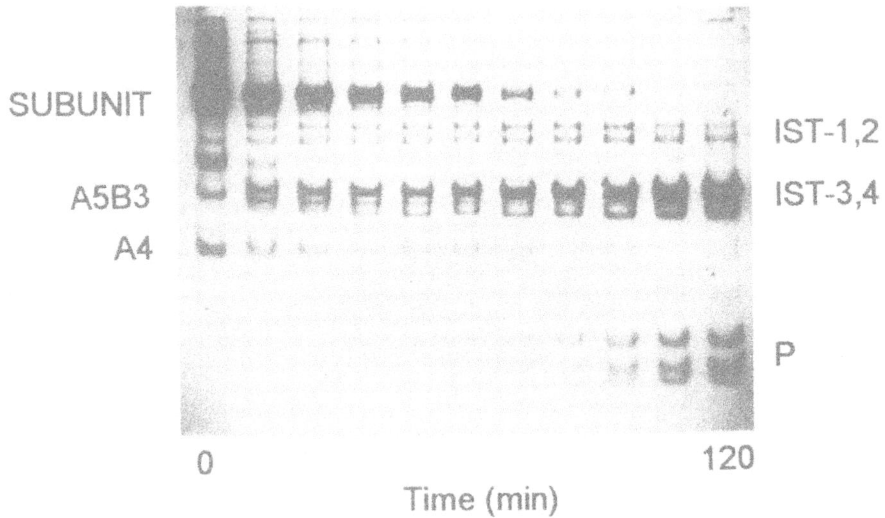

Figure 3. Time course analysis of fragmentation of glycinin-T by a SDS gel electrophoresis (adapted from Kamata et al., 1991a with permission.)

The proteolytic attack mainly occurs in the acidic polypeptide chains. This indicates that the acidic polypeptide chains exist in a highly susceptible region, probably, at the surface of the molecule and have labile regions for proteolysis.

We tried partial analysis of the conformational information on glycinin by the combined use of prediction methods for the secondary structure and hydrophobicity from amino acid sequences, secondary structure prediction from circular dichroism data, and the fragmentation analyses by limited proteolysis. Figure 4 (Kamata et al., 1991a) shows a conformational prediction from sequence data of a glycinin subunit. Basic polypeptide chain mainly consists of β-sheet structure and hydropathy index shows that these β-sheet regions are hydrophobic and are located in the interior of the molecule. Arrows show the digestion points predicted from the amino acid sequence analyses of the fragments. The basic chain has no digestion point, which suggests that this part of the subunit is in the interior of the molecule. The acidic polypeptide chain has hydrophilic regions, at the top regions of the hydropathy index curve; these are marked as I1, I2 and I3. It should be noted that the digestion points are located in these parts. Also, these parts are rich in β-turn structure and random structure. From these analyses, we presume that these parts are flexible and are located at surface of the subunit. Proteolytic enzymes can easily attack such parts of the molecule.

The P fragments associate with glycinin-T via noncovalent forces, unless these are dissociated by denaturants. The location of the P fragments was revealed by an N-terminal sequence analysis of the fragments; these fragments are generated between regions A and B. The attack site of trypsin in the A region is Lys or Arg at around residue number 100. The T fragments were on the N-terminal side, because these fragments were connected to the basic polypeptides by a disulfide bond at Cys86 (small arrow). The remainder of the acidic polypeptide chain after the B region must be digested in an area between the B region and the C-terminal of the acidic polypeptide chain, because the other detectable fragments were not found in SDS gel electrophoresis by which low-molecular-weight fragments below 10,000 are difficult to detect. Also, this area contained the I3 hydrophilic region and more Lys and Arg residues than the others; therefore, this part of the subunit would be easily degraded to small peptides. This area also contained a hypervariable region (Nielsen et al., 1989) that may have little importance in maintaining glycinin conformation. Therefore, it is possible that this area is not stacked compactly.

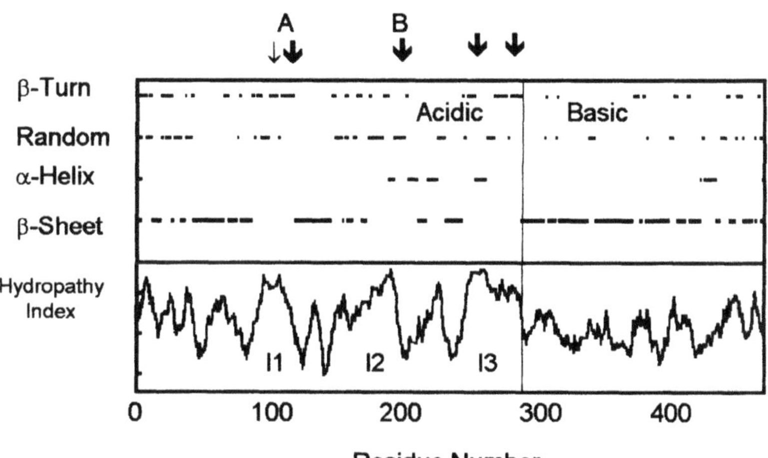

Figure 4. Secondary structure prediction from the amino acid sequence and hydropathy index analysis of a glycinin subunit (adapted from Kamata et al., 1991a with permission.)

It has been revealed that the A4 polypeptide chain and the A5B3 subunit were generated by a post-translational modification by proteolytic digestion from a precursor subunit (Nielsen et al., 1989; Momma et al., 1985). The location of the nicking site was near the A region of the precursor, and the flexibility of the region caused limited proteolysis. However, the regions of other subunits also seem to be flexible. The reason for proteolysis in vivo, therefore, is not only the flexibility of the protein conformation, but also the special amino acid sequence that fits the protease specificity for limited proteolysis. From these considerations, the A5B3 subunit may be rather similar to IST-4 and found in a similar position to the fragment on SDS gel at 0 min lane (Fig. 3).

The β-sheet regions were clearly separated by flexible regions. This may indicate that the protein subunits consist of some structural domains composed of β-sheet structures. The T fragments, P fragments and the basic polypeptide chains may be the domain structures. The methods used in these studies, i.e., secondary structure prediction from the amino acid sequence, hydropathy index analysis and the fragmentation pattern analysis with trypsin, reveal that the glycinin subunits possess complex structure features. Each of the methods is not effective alone, but a combination of these methods provides a more effective information on the structural features of glycinin.

Figure 5 shows a degradation scheme of a glycinin subunit. Glycinin is made of six subunits packed as shown in Fig.5 (top and side views). The basic polypeptide chain is located inside of the molecule. Initial enzyme attack releases some small peptides. Following attacks separate the acidic polypeptide chain into two parts, but these two parts do not disconnect from the whole molecule. Therefore, glycinin-T almost keeps its native structure.

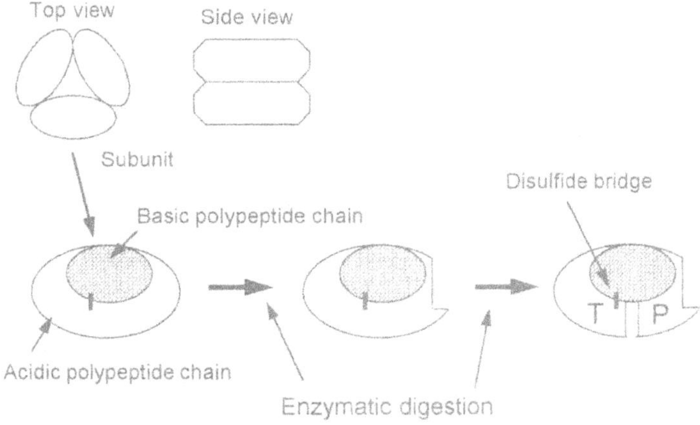

Figure 5. Schematic representation of glycinin digestion.

The structural information described here is thought to be useful for understanding the role in the functional properties of each domain in the subunit of soybean proteins. For example, the variations in the coagulation properties of soybean proteins have been studied in connection with the role of the fragments in disulfide network formation (Yamagishi et al., 1987). The information is also useful for predicting the effect of chemical, enzymatic and genetic modifications on the functional properties of soy glycinin.

FUNCTIONAL PROPERTIES OF GLYCININ-T

Emulsifying Properties of Glycinin-T

We studied the emulsifying properties of glycinin digests obtained by limited trypsinolysis (Ochiai et al., 1982; Kamata et al., 1984). Among the fractionated digests, the high molecular weight fragments of heat-denatured glycinin (HMF; open triangle) and the digest of native glycinin (DNG = Glycinin-T; open square) showed higher emulsifying properties than the intact glycinin (open circle). On the other hand, the precipitated fraction during the digestion of heat-denatured glycinin (PPT; filled triangle) and the low molecular weight fragments of heat-denatured glycinin (LMF; filled circle) poorly stabilized the emulsion (Fig. 6; Kamata et al., 1984).

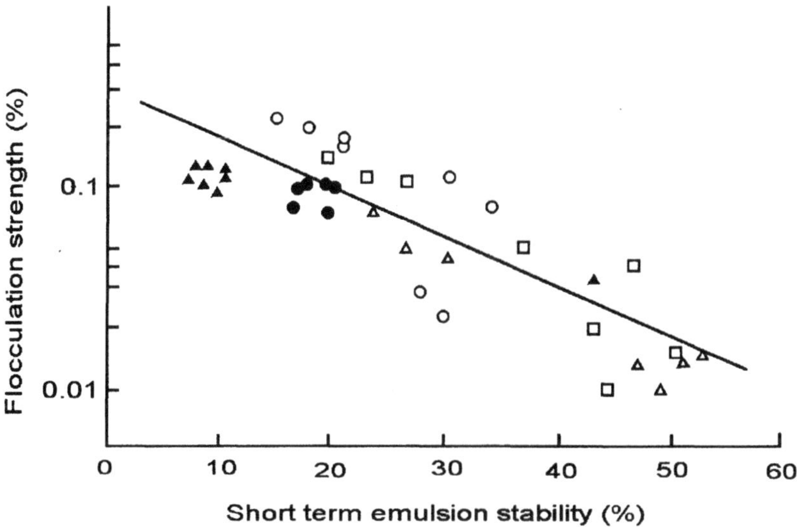

Figure 6. Relationship between the flocculation strength and the emulsion stability (from Kamata et al., 1984 with permission.)

A new parameter, flocculation strength, which represents the strength of holding force of flocculates among the emulsion particles, has been proposed. This was defined as the detergent concentration required to dissociate the oil-droplet flocculates. The relationship between the flocculation strength and the emulsion stability is shown in figure 6. The reasons for differences in the emulsifying properties were analyzed.

The sizes of the emulsion particles were examined. In spite of the differences in the emulsifying properties, there were no considerable size differences among the particles emulsified by the good emulsifiers. However, in cases where dissociation of the flocs did not occur in the presence of detergents, large size formed as determined by spectroturbidimetry. The flocculation is probably related with the emulsion properties, because it is thought that the flocculation is the pre-process of creaming. The emulsion particles stabilized by the relatively poor emulsifiers rapidly and strongly flocculated. The flocculation and the protein adsorption may be closely related to the stability of the emulsions stabilized by the soy protein digests. This is supported by the close relationship between the flocculation strength and the emulsion stability (Fig. 6)

Conformational analysis, as determined by urea-induced conformational transitions (Fig. 7; Kamata et al., 1984) and other methods, show that the best emulsifiers (HMF and DNG) are characterized by a fairly large molecular weights and a non-random conformation. For example, HMF and DNG, which exhibit better emulsifying properties (Fig. 6), still possess residual three dimensional structures as indicated by conformational transitions in urea solutions (Fig. 7). However, the structure seems to be more flexible than the native one, because changes in conformation begin at lower urea concentration. There is a possibility that a part of the internal hydrophobic region is exposed because of the flexibility of the proteins. The three steps involved in forming a protein-stabilized emulsion are: (1) diffusion of protein molecules or aggregates to and attachment at the interface; (2) spreading or unfolding of adsorbed molecules at the interface; (3) molecular rearrangement and reconformation of the adsorbed molecules. The protein solubility and surface hydrophobicity may facilitate the first step and the conformational flexibility may be required for the second and third stages.

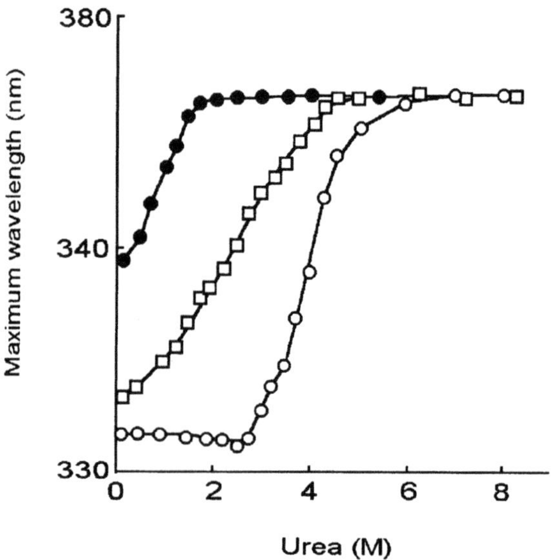

Figure 7. Denaturation curves of the digests. The changes of maximum wavelength of natural fluorescence of the protein were measured as an indication of the protein denaturation (from Kamata et al., 1984 with permission.)

Soy Protein Curd Formation by Limited Proteolysis and Soy Milk Cheese

Thermally-induced gels of soybean protein have high water holding capacity and elasticity (Kinsella, 1979). We tried to make the protein possess other characteristics by limited hydrolysis (Kamata et al., 1989). Native soy glycinin and β-conglycinin were partially hydrolyzed by trypsin. Texture profile analysis showed that the treated globulins had lower gel hardness values than the intact globulins. Moreover, the gels of the protease treated globulins exhibited brittleness, which was not observed with the intact globulin gel. The protease treated glycinin formed a turbid and rough gel that resembled tofu or cheese curd (Fig. 8a, b; Kamata et al., 1989), whereas the protease treated β-conglycinin formed a transparent and fragile (Fig. 8c, d).

Figures 8a and 8b show the gels made from glycinin and glycinin-T. Glycinin gel was a little transparent (Fig. 8a, left) but glycinin-T gel was completely opaque (Fig. 8a, right). When the gels were compressed to 80% (Fig. 8b), the glycinin gel was not crushed, whereas glycinin-T gel crumbled. The appearance of glycinin-T gel was rather like a coagulum than a gel. It looked like a cheese curd. Based on these observations, attempts were made to prepared cheese analogs using glycinin-T.

Soybean proteins have high nutritional value and various functional properties. Efforts to utilize this advantage of the protein to substitute of dairy foods, especially of cheese, have been made (Motoki et al., 1982; Nishiya et al., 1989; Park et al., 1985; Murata et al., 1987a; Murata et al., 1987b; Murata et al., 1989; Fuke and Matsuoka, 1987). Many workers tried to make soymilk-curd for making cheese analog by treatment with proteases (Murata et al., 1987a; Murata et al., 1987b; Murata et al., 1989; Fuke and Matsuoka, 1987).

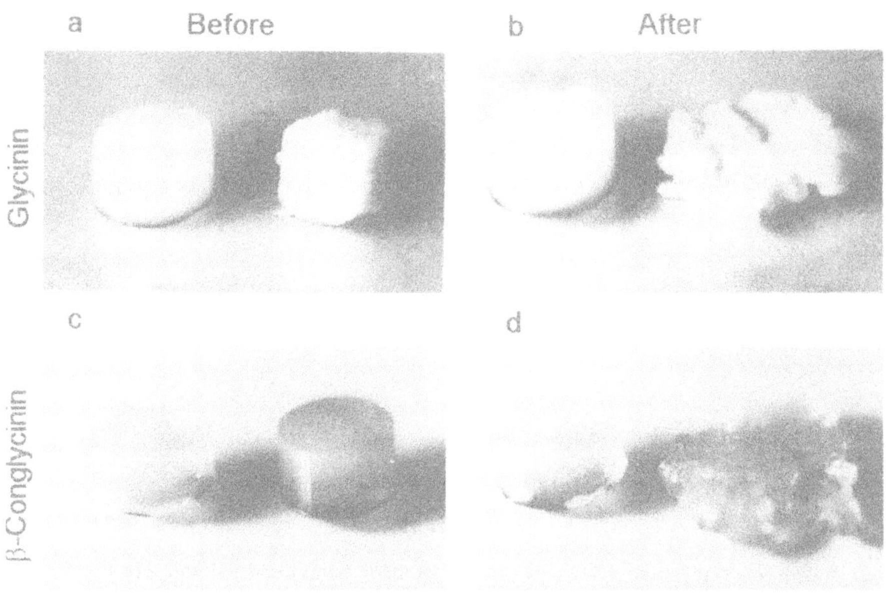

Figure 8. Heat induced gels of glycinin and β-conglycinin treated without (left) and with (right) trypsin before and after compression (adapted from Kamata et al., 1989 with permission.)

Work is being carried out in our laboratory to make a cheese analog from soymilk by limited proteolysis using an immobilized enzyme reactor system (Kamata et al., 1991b; Kamata et al., 1992a; Kamata et al., 1993). This system is effective for preserving enzymes, for constructing a continuously flowing process and for the controlling the degree of proteolysis. However, separation of the proteolysis and coagulation processes is necessary because the coagulated soy milk can plug the reactor. We reported that limited proteolysis with plant proteases (papain and bromelin) under low temperature condition (5; Kamata et al., 1992a), followed by heat treatment resulted in efficient clotting of soymilk. The conditions are suitable for an immobilized enzyme system. Therefore, it was concluded that the use of the plant proteases was suitable for a cheese analog production.

Rapid heating with a microwave oven was employed for both enzyme inactivation coagulation of soy protein. Because slow heating frequently caused loss of coagulability of soybean protein as a result of over-digestion (Kamata et al., 1991b). The use of a microwave oven was helpful to solve this problem. However, in an industrial-scale process this may be difficult to do and may also increase the cost. On the other hand, soy milk treated with immobilized enzyme can be coagulated by slow heating because no enzymatic activity would remain in the soy milk after treatment with the immbobilized enzyme reactor (Kamata et al., 1993). Therefore the use of an immobilized enzyme system is favorable.

The production of cheese analogs as fabricated foods without fermentation was attempted (Kamata et al., 1991b). In this case, a part of the soy milk-curd was vacuum-dried. A part of the dried sample was washed with hot water and dried again. These dried samples and wet soy milk-curd were used as starting materials. The subsequent steps were as follows. Soy milk-curd or dried soy milk-curd, sodium caseinate and other ingredients were added to solid fat melted at 80 °C (Fig. 9). The mixture was well emulsified with a kneader; however, the addition of flavors sometimes interfered with the emulsion formation. In that case, a high-speed mixer instead of the kneader is preferable.

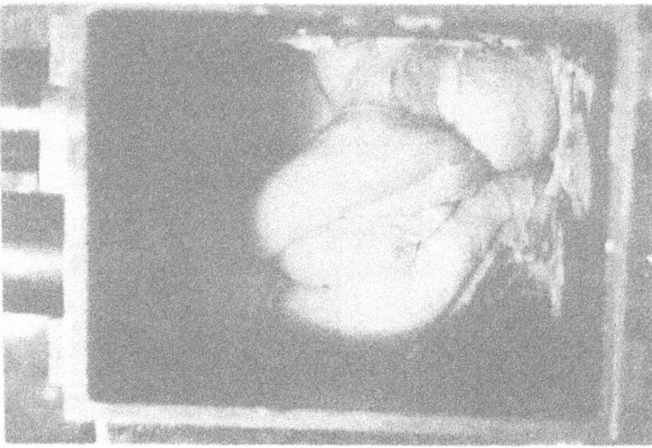

Figure 9. The kneading process of the soymilk cheese.

A typical composition of the cheese analogs made from the dried and wet soy milk curd are shown in Table 1 (Kamata et al., 1991b). The analog made from wet curd had good texture and acceptable taste. In this product, about 55% of the total protein was from soybean (Table 1), which is much higher than the 30% reported in previous similar products. It is possible to increase the percentage of soy protein up to 70% by using the dried soy milk-curd; however, the taste of the analog becomes undesirable. The concentration of undesirable-taste compounds, such as saponins, is thought to be a reason for the unacceptable taste. Therefore, we tried to wash off these compounds with hot water. The soybean-curd powder thus obtained had no taste and the analog made from the powder had acceptable taste and about 76% soybean protein. Although the washed curd powder had not

Table 1. Compounds of cheese analogs from soybean-curd[a]

Component	Weight (g)		
	Wet	Dried	Dried and Washed
Curd	71.3	44.1	44.1
Sodium caseinate	11.2	10.0	10.0
Hardened oil	25.7	22.7	22.7
Salt	1.5	1.5	1.5
Emulsifying salt	2.0	3.0	2.0
Flavor 1	5.0	5.0	50
Flavor 2	0.1	5.0	2.5
Water	10.0	66.0	76.0
Water content (%)	46.6	45.6	49.4
Protein share (%)[b]	55.3	71.1	76.0

[a]from Kamata et al., 1991b with permission
[b]Soybean protein proportion in whole protein

only good taste but also favorable white color, the yield of protein was quite low. Much of the soy protein must have been washed off during the washing step. Therefore, it was concluded that the wet soy milk-curd was the best material for preparing cheese analogs because of its good yield and better taste.

IMMOBILIZED ENZYME SYSTEM FOR FOOD PRODUCTION

Immobilization of Enzymes by Enzymatic Crosslinking Using Transglutaminase

We attempted to use an immobilized enzyme system for fabricating cheese analog. Chemical reactions are usually required to immobilize enzymes (Chibata,1975), and the chemicals used for immobilization must not contaminate an industrial product, especially foods or drugs. Therefore, after immobilization, the system must be exhaustively washed to remove any hazardous chemical residues that might be present. This requirement is tedious, costly, and may have an adverse effect on enzyme activity. Enzymatic crosslinking using enzymes such as transglutaminase (EC 2.3.2.13) is one of the solution to the problem and this has been tried previously. Motoki et al. (1987) immobilized some enzymes by entrapping in protein gel network obtained by the transglutaminase reaction. However, this method is not suitable for macromolecular substrates because the substrates like soyprotein in soy milk never enter the gel network. A better method is needed for using transglutaminase in enzyme immobilization for macromolecules.

It has been reported that enzymes adsorbed on ion exchangers can be immobilized by polymerizing the adsorbed enzyme molecules via crosslinking; the polymerized enzyme film forms a coat on ion exchange support beads, but are not crosslinked to the beads (Kurota et al., 1990) (Figure 10). Microbial transglutaminase (MTG; Ando et al., 1989) was used instead of chemical crosslinkers (Kamata et al., 1992b).

The effects of pH and the degree of added MTG activity during immobilization of trypsin on SP-Sephadex C-25 are shown in Figure 11 (Kamata et al., 1992b). The activity of immobilized trypsin increased with increasing MTG activity (0, 83 and 467 nkat). The immobilized trypsin exhibited maximum activity in the pH range 6-10.

Although more trypsin could be adsorbed to the ion exchanger at low pH because of an increase in net positive charge of trypsin at low pH, trypsin could not be immobilized at pH 2~3. This is because the optimum pH of MTG is 6-7 (Ando et al., 1989) and MTG had little activity at pH 2-3. Further, trypsin did not adsorb to the cation exchanger above pH 10

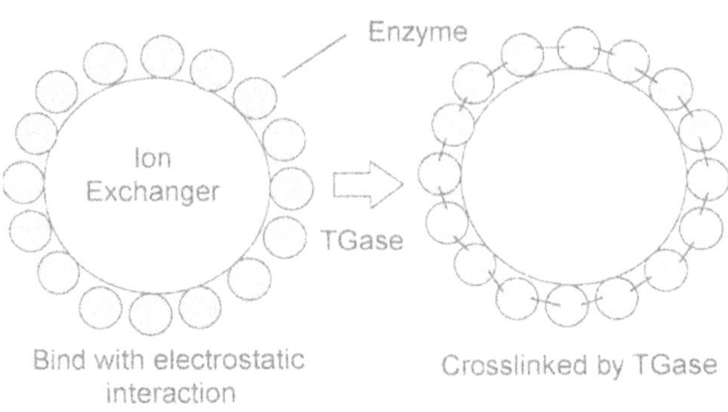

Figure 10. Enzyme immobilization on an ion exchanger with microbial transglutaminase.

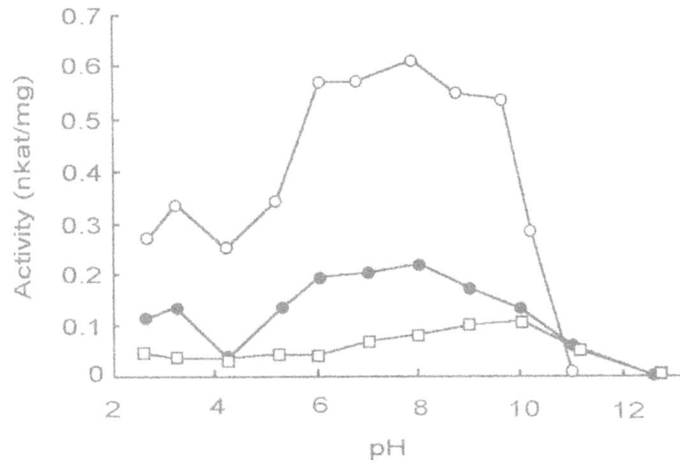

Figure 11. Effect of microbial transglutaminase (MTG) concentration on immobilized tryptic activity (adopted from Kamata et al., 1992b with permission.)

because it had a net negative charge above its isoelectric point (pI = 10.5). On the other hand, it was possible to immobilize α-amylase on QAE-Sephadex at pH 9-11, and on SP-Sephadex at pH 4 using MTG. These results indicate that the net charge of the enzyme at a given pH affects its adsorption to ion exchangers. Further, these results also clearly demonstrate that MTG can be successfully used to immobilize enzymes on ion exchangers.

The above method of enzyme immobilization has many advantages for use in food and drug production. The immobilizing process is also very simple and mild. An enzyme is adsorbed on an appropriate ion exchanger and then the support is incubated with MTG. The immobilized enzyme system also can be expected to be safe, because no hazardous chemicals is being used.

Enzyme Immobilization on Glycosylated Egg White Beads

Another way of immobilizing enzyme is the use of edible proteins as the support material. In browning reaction, proteins react with sugars and form advanced glycosylated end products (AGE). The AGE product can react to form intermolecular crosslinkages with other proteins such as enzymes (Monnier et al., 1984; Okitani et al., 1984; Kato et al., 1987; Pongor et al., 1984). We used this phenomenon for enzyme immobilization (Kamata et al., 1990). For this purpose, we used egg white beads as a support protein.

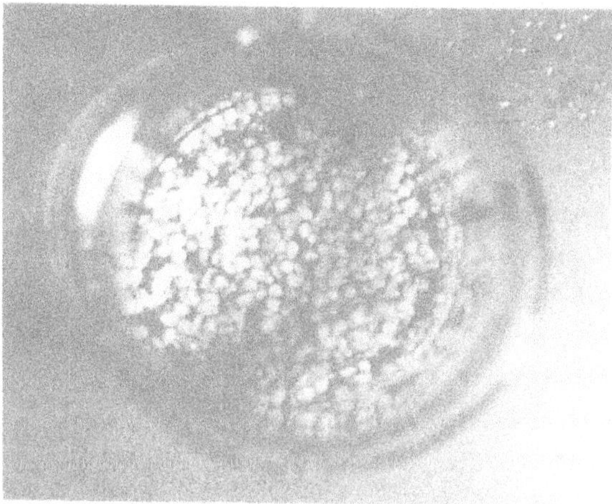

Figure 12. Egg white beads for enzyme immobilization.

Egg white beads were prepared by using a method similar to that used for preparing immobilized microorganisms. We used sodium alginate as a gelling agent. The egg white and sodium alginate mixture was dropped into a calcium chloride solution. This resulted in the formation of gel beads, 2-3 mm in diameter. The beads were boiled to coagulate the egg white. The beads were then incubated in 20% glucose solution at 80 for 15 hr. Browning reaction proceeds under these conditions (Fig. 12). The beads were washed with excess water, then lyophilized. To prepare an immobilized enzyme, the beads were added to enzyme solution and then incubated at 30-40 °C for 24 hrs. The immobilized enzyme was used after washing with excess water.

Figure 13 shows a comparison of the operational stability of trypsin immobilized on various supports. Trypsin was immobilized by the standard methods for each support. The substrate was 1% casein solution. Similar amounts of the activity were set in a continuous stirring type reactor (4 ml volume). One of the most popular support is chitosan beads, but rapid decrease of trypsin activity with elution volume indicates that this support is not suitable for protein digestion.

The activity per dry weight of trypsin immobilized on various supports was as follows: egg-white beads (272 nkat/g), SP-Sephadex C-25 (352 nkat/g), Chitopearl BCW 3010 (94 nkat/g) and CNBr-Sepharose 4B (68 nkat/g). A high activity per dry weight was obtained in the case of egg-white beads. Enzyme stability was better in the case of CNBr-Sepharose, but this support is extremely expensive and use of a cyanide to crosslink the enzyme may not be undesirable for food processing purposes. On the other hand, egg white beads show better results. This support is inexpensive and also edible because no toxic chemicals are used in its preparation. The commercial supports listed in Fig. 13 are about 80~700 fold more expensive than egg white beads.

One of the major problems encountered in the use of immobilized enzyme reactors for food processing is the inactivation of the enzyme during processing. For example, several workers (Ohmiya et al., 1979; Angelo and Shahani, 1983; Mashaly et al., 1988; Carlson et al., 1986; Garg and Johri, 1993) have reported rapid inactivation of immobilized rennet or chymosin during cheese curd making. Also, the lifetime of an immobilized enzyme system for making cheese analogs from soy milk was only a few hours (Kamata et al., 1993). It seems that this problem is common in immobilized enzyme systems that use macromolecular substrates, such as proteins.

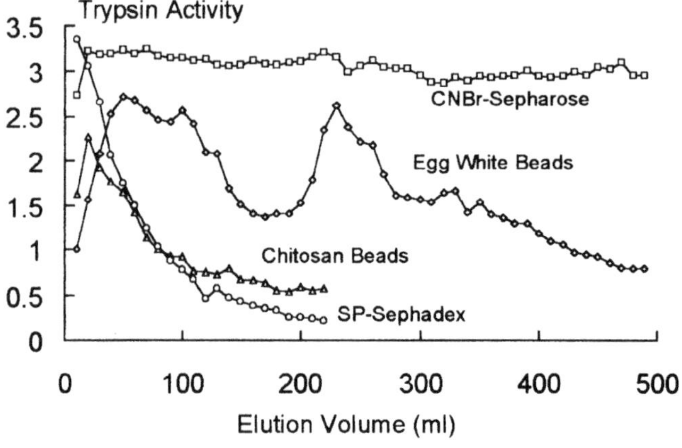

Figure 13. Comparison of the activity and stability of various immobilized enzyme.

The main reason for the rapid inactivation of enzymes immobilized on the hard supports like the chitosan beads and ion exchangers may be adhesion of the substrate protein or the digested peptide on the surface of the immobilized enzyme. The adhered materials may interfere with the access of the substrate to the immobilized enzyme. Washing of the adhered substrate with water was found to be effective to some extent on recovery of the activity (Kamata et al., 1993). Soft supports, such as egg-white beads gradually collapsed, possibly because of proteolysis.

The lifetime of an immobilized enzyme systems in food processing must not exceed a few days because of microbial contamination. Because it is often impossible to sterilize an immobilized enzyme and a substrate in a reactor system, the immobilized enzyme must be discarded and the whole system must be completely cleaned up regularly. Therefore, the carrier cost must be as low as possible.

The typical operation of an immobilized enzyme reactor in food industry involves repetition of following steps: (i) applying the immobilized enzyme to the system; (ii) operate for a few days; (iii) discard the immobilized enzyme and (iv) sanitize the whole system.

PROTEIN-CHITOSAN HYBRID FOR FOOD USE

Covalent attachment of a polymer to a protein molecule has been considered as a useful method to improve its biological activities and functional properties (Inada, 1987). In practical applications, however, safety of these polymers should be taken into consideration. Therefore, an attempt to find and characterize new polymers of which biological safety has been established seems to be important. Among various candidate polymers, carbohydrates have been found to be suitable for food use (Baniel et al., 1992; Nakamura et al., 1992a; Nakamura et al., 1992b; Kato et al., 1993; Kato, 1994). In this context, crosslinking of polysaccharides to proteins has been attempted by using the Maillard reaction (Nakamura et al., 1992a; Nakamura et al., 1992b; Kato et al., 1993; Kato, 1994).

Chitosan is one of the excellent candidates for this purpose. Chitosan and its oligomers have antibacterial, antifungal, and antitumor activities (Yamasaki et al., 1992). They are used in foods, medicines, and cosmetics. Also they have plenty of positive charges that will affect the biological and functional properties of a chitosan-protein hybrid. Therefore, crosslining of this polysaccharide to proteins may improve functionality of proteins. However, chitosan is not soluble at around neutral pH, which makes it difficult to use this polysaccharide in food systems. Therefore, producing soluble chitosan is important for food applications.

Chitosan is soluble in dilute acid solution but not in water at neutral pH. On the other hand, chitosan oligomers are soluble in water. Therefore, we tried to make chitosan soluble by limited hydrolysis using chitosanase. One gram of the chitosan was dissolved in 100 ml of the 2% acetic acid solution. Ten milligrams of a chitosanase was added and incubated for 24 hr at room temperature. Then the pH of the solution was adjusted to 6.0, which is the optimum pH of the enzyme. An extra amount of the enzyme (10 mg) and sodium azide (0.2 g) were added and incubated for another 24 hr at room temperature. The pH of the mixture was adjusted to 8.0 and the precipitate formed was removed by centrifugation (3,000 rpm, 10 min). The supernatant was dialyzed against water at 5 °C and lyophilized. The powder was named as water soluble chitosan (WSC) and stored at -20. The yield of WSC was about 30%. This preparation was soluble in water at neutral pH but may not contain chitosan oligomers because these have been removed during dialysis.

Creation of intermolecular bonding between a WSC molecule and a protein molecule was attempted by using the Maillard reaction. We selected α_{s1}-casein as a target protein. Equal amounts (4 mg/ml) of α_{s1}-casein and WSC were dissolved in a potassium phosphate buffer (pH 7.8). The mixture was heated at 80 °C for 15 hr in a screw capped tube. Figure 14 shows Sepharose CL-6B chromatography of the mixture. The eluate was monitored at 280 nm and the fractions were analyzed by the indole-HCl method for detecting chitosan. Figure 14a shows the elution patterns of casein-WSC mixture (2 ml) monitored at 280 nm with or without the heat treatment. Before heat treatment (open circle), only the ultraviolet absorption of the protein can be seen. On the other hand, the absorption of chitosan becomes detectable in the low molecular weight region of the chromatogram (filled circle) because a part of the chitosan has undergone browning reaction, the products of which absorb UV light.

Figure 14b shows the elution profiles of chitosan detected by the indole-HCl method. WSC in the non-heated mixture eluted at the low molecular weight region (open circle) suggesting no interaction with casein. After the heat treatment, a part of WSC shifted toward the casein peak (filled circle), which indicated the formation of a chitosan-casein complex.

These results clearly show that the heat treatment caused interaction between WSC and α_{s1}-casein. However, the bonding forces involved in these complexes are not clear yet. Disappearance of the casein band was observed on a SDS gel electrophoresis. At the same time, high molecular weight substances also appear, suggesting the formation of covalent bonds between chitosan and α_{s1}-casein molecules (Fig. 15).

Some preliminary experiments were done on the change in protein functionality by the reaction with WSC. Soybean glycinin heated with WSC showed higher viscosity at a certain protein concentration (10%) than that without WSC. On the other hand, the viscosity of 12% glycinin solution heated with WSC was fairly lower than without WSC. This phenomenon suggests that WSC produces intermolecular crosslinkage among soy proteins, especially at high protein concentration. At lower protein concentration, incorporation of chitosan and some crosslinking of the protein may increase the molecular weight of the protein and thereby increase the viscosity. On the other hand, covalent gel networks may be developed at higher protein concentration. This network should be destroyed by the mechanical flow in viscosity measurement with a double-cylindrical rotation-viscometer.

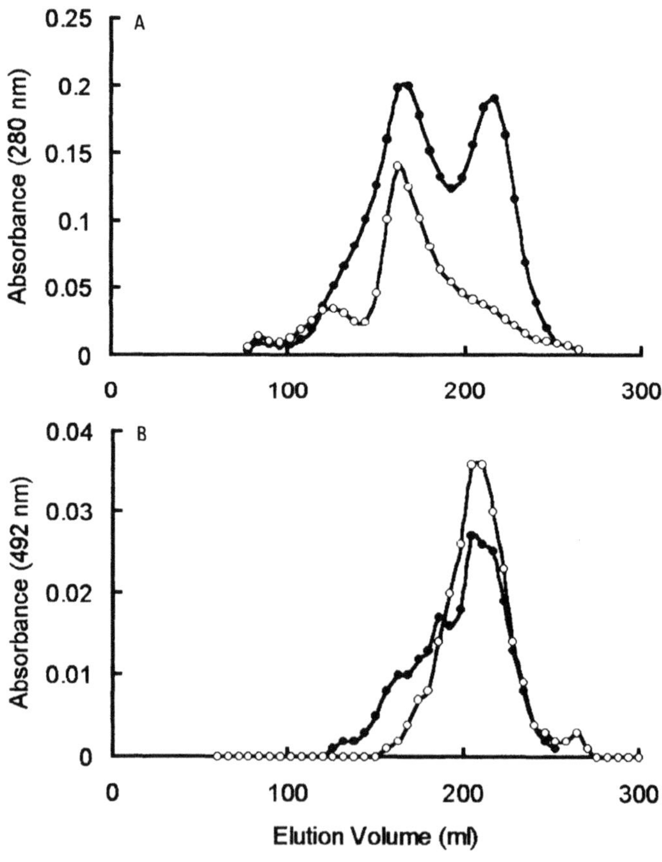

Figure 14. Elution profiles of α_{s1}-casein-WSC mixture on Sepharose CL-6B with or without heat treatment.

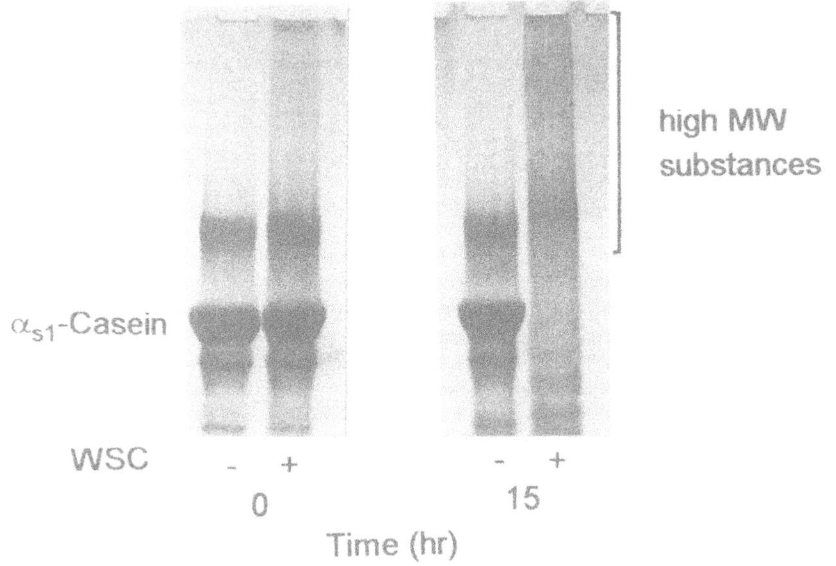

Figure 15. SDS polyacrylamide gel electrophoresis of α_{s1}-casein heated with WSC.

We are now attempting the introduction of WSC to lysozyme for improving the antimicrobial activity. In these experiments, we found that the crosslinks are reversible and are ruptured with time. Reduction of the Schiff's base produced by the Maillard reaction is one of the ways to stabilize the linkage. We are trying to use an electro-reduction system for this purpose because we cannot use reducing agents in food applications. The usefulness of the protein-chitosan hybrid is not confirmed as of yet, however, the additional positive charges must change and improve the functional properties of proteins.

REFERENCES

Ando, H., Adachi, M., Umeda, K., Matsuura, A., Nonaka, M., Uchio, R., Tanaka, H., and Motoki, M., 1989, Purification and characteristics of a novel transglutaminase derived from microorganisms, *Agric. Biol. Chem.* 53:2613.

Angelo, I. A., and Shahani, K. M., 1983, Coagulation of milk with rennet immobilized on Sepharose-4B, *Ind. J. of Dairy Sci.* 36:45.

Baniel, A., Caer, D., Colas, B., and Gueguen, J., 1992, Functional properties of glycosylated derivatives of the 11S storage protein from pea (*Pisum sativum* L.), 40:200.

Carlson, A., Hill, G. C., and Olson, N. F., 1986, The coagulation of milk with immobilized enzymes: a critical review, *Enzyme Microb. Technol.* 8:642.

Chibata, I., 1975, Methods in enzyme and cell immobilization, in: *Koteika Kouso (Immobilized Enzyme)*, I. Chibata, ed., Kodansha, Tokyo.

Cowgill, R. W., 1975, Proteolysis of paramyosin from *Mercenaria mercenaria* and properties of its most stable segment, *Biochemistry* 14:503.

Fuke, Y., and Matsuoka, H., 1987, Changes in proteins and protease activities during ripening of cheese like product from soymilk using penicillium caseicolum, *Nippon Shokuhin Kogyo Gakkaishi* 34:826.

Garg, S. K., and Johri, B. N., 1993, Immobilization of milk-clotting proteases, *World J. Microb. Biotech.* 9:139.

Inada, Y., 1987, *Tanpakushitsu Haiburiddo (Protein Hybrid)*, Kyoritsu Shuppan, Tokyo.

Kamata, Y., and Shibasaki, K., 1978a, Degradation sequence of glycinin by tryptic hydrolysis, *Agric. Biol. Chem.* 42:2103.

Kamata, Y., and Shibasaki, K., 1978b, Formation of digestion intermediate of glycinin, *Agric. Biol. Chem.* 42:2323.

Kamata, Y., Okubo, K., and Shibasaki, K., 1979a, Decrease of the soybean glycinin digestibility in excess denaturation; Effect of refolding, *Agric. Biol. Chem.* 43:1219.

Kamata, Y., Kimigafukuro, J., and Shibasaki, K., 1979b, Limited tryptic degradation of urea denatured glycinin, *Agric. Biol. Chem.* 43:1817.

Kamata, Y., Kikuchi, M., and Shibasaki, K., 1980, Non-dissociative small fragments of glycinin-T, *Agric. Biol. Chem.* 44:575.

Kamata, Y., Otsuka, S., Sato, M., and Shibasaki, K., 1982, Limited proteolysis of soybean -conglycinin, *Agric. Biol. Chem.* 46:2829.

Kamata, Y., Ochiai, K., and Yamauchi, F., 1984, Relationship between the properties of emulsion systems stabilized by soy protein digests and the protein conformation, *Agric. Biol. Chem.* 48:1147.

Kamata, Y., Takahata, H., and Yamauchi, F., 1989, Gelling properties of cleaved soybean globulins by limited proteolysis, *Nippon Shokuhin Kogyo Gakkaishi* 36:557.

Kamata, Y., Kurota, A., and Yamauchi, F., 1990, Enzyme immobilization on glycosylated edible protein, *Agric. Biol. Chem.* 54:3049.

Kamata, Y., Fukuda, M., Sone, H., and Yamauchi, F., 1991a, Relationship between limited proteolysis of glycinin and its conformation, *Agric. Biol. Chem.* 55:149.

Kamata, Y., Ohta, K., Yamauchi, F., and Yamada, M., 1991b, Cheese analogs from soybeans or soymilk-curd by limited proteolysis, *Nippon Shokuhin Kogyo Gakkaishi* 38:1143.

Kamata, Y., Chiba, K., Yamauchi, F., and Yamada, M., 1992a, Selection of Commercial enzymes for soymilk-curd production by limited proteolysis with immobilized enzyme reactor, *Nippon Shokuhin Kogyo Gakkaishi,* 39:102.

Kamata, Y., Ishikawa, E., and Motoki, M, 1992b, Enzyme immobilization on ion exchangers by forming an enzyme coating with transglutaminase as a crosslinker, *Biosci. Biotech. Biochem.* 56:1323.

Kamata, Y., Sanpei, S., Yamauchi, F., and Yamada, M., 1993, An immobilized protease reactor for cheese analog production from soymilk, *Nippon Shokuhin Kogyo Gakkaishi* 40:296.

Kato, A., Minaki, K., and Kobayashi, K., 1993, Improvement of emulsifying properties of egg white proteins by the attachment of polysaccharide through Maillard reaction in a dry state, *J. Agric. Food Chem.* 41:540.

Kato, A., 1994, New functional food proteins by polysaccharide modification, *Nippon Shokuhin Kogyo Gakkaishi* 41:304.

Kato, H., Shin, D. B., and Hayase, F., 1987, 3-Deoxyglucosone crosslinks proteins under physiological conditions, *Agric. Biol. Chem.* 51:2009.

Kinsella, J. E., 1979, Functional properties of soy proteins, *J. Am. Oil Chem.* Soc. 56:242.

Kitamura, K., Takagi, T., and Shibasaki, K., 1976, Subunit structure of soybean 11S globulin, *Agric. Biol. Chem.* 40:1837.

Koshiyama, I., and Fukushima, D., 1976, Physico-chemical studies on the 11S globulin in soybean seeds: Size and shape determination of the molecule, *Int. J. Peptide Protein Res.* 8:283.

Kurota, A., Kamata, Y., and Yamauchi, F., 1990, Enzyme immobilization by the formation of enzyme coating on small pore-size ion-exchangers, *Agric. Biol. Chem.* 54:1557.

Marco, Y. A., Thanh, V. H., Tumer, N. E., Scallon, B. J., and Nielsen, N. C., 1984, Cloning and structural analysis of DNA encoding an A2B1a subunit of glycinin, *J. Biol. Chem.* 259:13436.

Mashaly, R. I., Saad, M. H., El-Abassy, F., and Wahba, A. A., 1988, Coagulation of milk by calf rennet, pepsin and Mucor miehei rennet immobilized on large agarose beads, *Milchwissenschaft* 43:79.

Mihalyi, E. and Harrington, W. F., 1959, Studies on the tryptic digestion of myosin, *Biochim. Biophys. Acta* 36:447.

Momma, T., Negoro, T., Hirano, H., Matsumoto, A., Udaka, K., and Fukazawa, C., 1985, Glycinin A5A4B3 mRNA: cDNA cloning and nucleotide sequencing of a splitting storage protein subunit of soybean, *Eur. J. Biochem.* 149:491.

Monnier, V. M., Kohn, R. R., and Cerami, A., 1984, Accelerated age-related browning of human collagen in diabetes mellitus, *Proc. Natl. Acad. Sci. USA* 81:583.

Motoki, M., Torres, J. A., and Karel, M., 1982, Development and stability of intermediate moisture cheese analogs from isolated soybean proteins, *J. Food Process. Preserv.* 6:41.

Motoki, M., Aso, H., Seguro, K., and Nio, N., 1987, Immobilization of enzymes in protein films prepared using transglutaminase, *Agric. Biol. Chem.* 51:997.

Murata, K., Kusakabe, I., Kobayashi, H., Akaike, M., Park, Y. W., and Murakami, K., 1987a, Studies on the coagulation of soymilk-protein by commercial proteinases, *Agric. Biol. Chem.* 51:385.

Murata, K., Kusakabe, I., Kobayashi, H., Kiuchi, H., and Murakami, K., 1987b, Selection of commercial enzymes suitable for making soymilk-curd, *Agric. Biol. Chem.* 51:2929.

Murata, K., Kobayashi, H., Kusakabe, I., Teramoto, H., and Murakami, K., 1989, Preparation of fermented soymilk curd with commercial proteinases, *Nippon Shokuhin Kogyo Gakkaishi* 36:417.

Nakamura, S., Kato, A., and Kobayashi, K., 1992a, Bifunctional lysozyme-galactomannan conjugate having excellent emulsifying properties and bactericidal effect, *J. Agric. Food Chem.* 40:735.

Nakamura, S., Kato, A., and Kobayashi, K., 1992b, Enhanced antioxidative effect of ovalbumin due to covalent binding of polysaccharides, *J. Agric. Food Chem.* 40:2033.

Nielsen, N. C., Dickinson, C. D., Cho, T. J., Thanh, V. H., Scallon, B. J., Fischer, R. L., Sims, T. L., Drews, G. N., and Goldberg, R. B., 1989, Characterization of the glycinin gene family in soybean, *Plant Cell* 1:313.

Nishiya, T., Tatsumi, K., Ido, K., Tamaki, K., and Hanawa, N., 1989, Functional properties of imitation mozzarella cheeses containing soy protein and casein, *Nippon Shokuhin Kogyo Gakkaishi* 36:805.

Ochiai, K., Kamata, Y., and Shibasaki, K., 1982, Effect of tryptic digestion on emulsifying properties of soy protein, *Agric. Biol. Chem.* 46:91.

Ohmiya, K., Tanimura, S., Kobayashi, T., and Shimizu, S., 1979, Application of immobilized alkaline protease to cheese-making, *J. Food Sci.* 44:1584.

Okitani, A., Cho, R. K., and Kato, H., 1984, Polymerization of lysozyme and impairment of its amino-acid residues caused by reaction with glucose, *Agric. Biol. Chem.* 48:1801.

Park, Y. W., Kusakabe, I., Kobayashi, H., and Murakami, K., 1985, Production and properties of a soymilk-clotting enzymes system from a microorganism, *Agric. Biol. Chem.* 49:3215.

Pongor, S., Ulrich, P. C., Bencsath, F. A., and Cerami, A., 1984, Aging of proteins - Isolation and identification of a fluorescent chromophore from the reaction of polypeptides with glucose, *Proc. Natl. Acad. Sci. USA* 81:2684.

Staswick, P. E., Hermodson, M. A., and Nielsen, N. C., 1984, The amino acid sequence of the A2B1a subunit of glycinin, *J. Biol. Chem.* 259:13424.

Yamagishi, T., Takahashi, N., and Yamauchi, F., 1987, Covalent polymerization of acidic subunits on heat-induced gelation of soybean glycinin, *Cereal Chem.*, 64:207.

Yamasaki, Y., Fukumoto, I., Kumagai, N., Ohta, Y., Nakagawa, T., Kawamukai, M., and Matsuda, H., 1992, Continuous chitosan hydrolyzate production by immobilized chitosanolytic enzyme from enterobacter sp. G-1, *Biosci. Biotech. Biochem.* 56:1546.

A REVIEW OF THE INTERACTIONS BETWEEN

MILK PROTEINS AND DAIRY FLAVOR COMPOUNDS

A. P. Hansen

Professor of Food Science
North Carolina State University
Raleigh, North Carolina 27695-7624

ABSTRACT

The effect of sodium caseinate and whey protein concentrate on vanillin, benzaldehyde, citral, and d-limonene was determined by quantitative descriptive analysis deviation from reference. A trained taste panel evaluated samples containing a single flavor compound in 2.5% sucrose solution against a reference sample. Vanillin, benzaldehyde, and d-limonene flavor intensity decreased as the concentratin of whey protein concentrate increased. In a separate study, the ability of delipidated methyl ketones to bind straight and branched chain methyl ketones was determined. The concentration of straight chain methyl ketones bound by the milk protein powder was inversely proportional to the size of the ligand. Branched chain methyl ketones did not exhibit a trend in binding based on ligand size.

INTRODUCTION

Consumer acceptance of dairy products is often based on perceived nutrition. Additionally, consumers demand that the flavor of the dairy products be enjoyable. In an effort to enhance the nutrition of dairy products, the dairy industry manufactures numerous reduced fat products. These reduced fat products offer the consumer the ability to enjoy dairy products while monitoring their intake of lipid materials. However, milk fat is the reservoir for much of the native flavor compounds found in milk. When quantity of milk fat in a product is reduced or removed, the flavor profile of the product is altered. The protein content can be increased by as much as 25% to replace the fat. The proteins in dairy products play a more prominent role in determining the flavor profile of the dairy products.

The use of trade names in this publication does not imply endorsement by the North Carolina Agricultural Research Service, nor criticism or similar ones not mentioned.

Flavor perception by the consumer involves the combination of smell and taste[15]. Taste refers to the action where food particles are mixed with salivary enzymes to promote the release of flavor compounds which then interact with flavor receptors on the tongue[2,23]. Flavor is one of the most important attributes of a food product because it often determines whether a food is rejected or accepted. The flavors in dairy products result from the complex and careful balancing of ingredients and flavorings. If this harmonious balance changes, the character of the flavor is altered and, consequently, the palatability may be affected.

Milk proteins are often added to frozen dairy desserts to impart smoothness, whippability, and to prevent weak body and coarse texture. As the protein content of food is increased to compensate for the reduction of fat, the potential exists for the reduction of flavor intensity due to flavor compound interactions with proteins. Because flavor compounds give flavor impact at extremely low concentrations, even a small degree of interaction between flavor compounds and ingredients or packaging material can lead to flavor changes.

Beta-lactoglobulin, an 18,000 molecular weight globular protein, is the major protein found in bovine milk whey. Beta-lactoglobulin undergoes strong association with many flavor compounds, and depending upon its concentration, could conceivably affect the perceived flavor of formulated foods. Other "superfamily" proteins with similar binding affinities include RBP, bilin binding protein, and insecticyanin[4]. "Superfamily" proteins have a hydrophobic pocket which is accessible from the exterior of the protein and is believed to be the primary binding site for a variety of non-polar molecules[26]. In addition to the primary binding site, β-lg is thought to contain other hydrophobic areas capable of undergoing interactions with apolar molecules[19,24,25,28,34,35].

In fluid reduced fat dairy products the interaction between milk proteins and added flavor compounds may result in reduced perception of the flavor compounds[3,7,9,10,16,18]. Conversely, nonfat dry milk powder frequently absorbs flavors during storage[6,1,8,17,22,25]. The degree and type of flavor compound binding to the milk powder may adversely affect the aroma and taste of food products made from these milk proteins.

This paper will review flavor compound interactions in two types of dairy products. The first section will review the influence of sodium caseinate and whey protein concentrate upon the perceptions of vanillin, benzaldehyde, citral and *d*-limonene. The second section will examine the binding of methyl ketones to milk protein powder.

MATERIALS and METHODS

Materials: Perception of Flavor Compounds in the Presence of Milk Proteins

Sodium Caseinate (CAS) and whey protein concentrate (WPC) was obtained from New Zealand Milk Products (Petaluma, CA). Proximate analyses furnished by the manufacturer indicated that CAS (Alanate 180) contained 91.1% protein, 3.5% ash, 4.0% moisture, 1.1% fat, and .1% lactose and gave a pH of 6.6 in 5% aqueous solution at 21°C. The WPC (Alacen 855) contained 76.5% protein, 3.5% ash, 4% moisture, and 12.5% lactose and gave a pH of 6.7 in 5% aqueous solution at 20°C. Food-grade vanillin was obtained from Rhone Poulenc, Princeton, NJ. Food-grade benzaldehyde, *d*-limonene, and citral were obtained from Mother Murphy's Flavors (Greensboro, NC).

The samples were prepared and the sensory panelists were trained as described in Hansen and Heinis[13,12]. All solutions were prepared in a 2.5% sucrose solution. CAS and WPC concentrations were 0%, .125%, .25%, and .5%. The concentrations of flavor compounds were vanillin (78.5 ppm), benzaldehyde (10, 20, and 30 ppm), citral (5, 15, and 20 ppm), and *d*-limonene (30, 60, and 90 ppm). The samples were held for 17 hrs at 6°C to simulate ice cream aging and to allow the protein-flavor compound interactions to equilibrate. The samples were allowed to come to ambient temperature (23°C). Samples were presented to the panelists in individual booths illuminated with red light. The panelists were asked to rank the flavor intensity of each sample as compared to a reference sample on the samples on the basis of a specific flavor. The four references consisted of a 2.5% sucrose solution containing a single flavor compound: vanillin (78.8 ppm), benzaldehyde (17.8 ppm), citral (19.8 ppm), or *d*-limonene (53.0 ppm). After the samples were rated, the individual flavor and aroma notes were noted, and samples were retasted and discussed by the panel. Results were analyzed by PROC GLM of SAS[30], and means were compared by the Duncan-Waller procedure.

Materials: Binding of Methyl Ketones to Milk Protein Powder

Fresh raw skim milk was obtained from the Cornell University dairy herd. The raw skim milk was treated with chilled acetone to precipitate the milk protein[14]. The delipidated protein was dried in a vacuum chamber for three days at 20°C. The dry protein was then sieved through a 120 mesh screen for uniform particle size. The protein powder was packed in a glass column, 3.8 cm (OD) x 127.0 cm. The column was packed in the following order (from the nitrogen tank): silanized glass wool, drying agent (desiccant), silanized glass wool, delipidated milk protein powder, and silanized glass wool. Nitrogen gas was purged through the glass column for four days to remove any acetone or volatile flavor compounds.

Surface area of the dried delipidated milk protein powder was determined by a B.E.T. surface analyzer, Quantasorb sorption system, (Quantachrome Corp., Syosset, NY) by measuring the change in thermal conductivity of a gas mixture, 10% nitrogen and 90% helium, as it passed over .2 g lots of milk protein powder.

Preparation of Sample Column. Borosilicate glass samples tubes, 6 mm ID x 100 mm, were packed with .2 g of silanized glass wool and .1 g protein powder. About half of the glass wool was placed in the base of the sample tube, then the protein powder was filled into the sample tube while vibrating the tube continuously to achieve uniform packing. The remaining glass wool was used to pack down the powder. Graphite grease was placed at the outside of both ends of the sample tubes to form an air tight seal when the graphite ferrules were tightened down on the glass tubing.

Gas Chromatography. A Hewlett Packard 5880A gas chromatograph (GC) (Hewlett Packard, Avondale, PA) with a Silar 10C (Supelco, Inc., Bellfonte, PA) and a modified ECID analyzer (Scientific Instrument Service, River Ridge, LA) was used as reported by Hansen et al.[11]. Before placing the packed sample tube into the ECID inlet, 1 µl of a methyl ketone was injected into the center of the glass wool at the top of the sample tube. The tube was placed in the ECID inlet and heated to 50°C. Heating volatilized the methyl ketone and allowed it to bind to the protein powder while avoiding heat denaturation of the milk protein. The tube was flushed for 20 s with carrier gas to force the volatile ketones through the milk protein powder. The sample tube was flushed at 0, 150, 300, 450, 600,

750, 1350, 1500 and 1650 seconds after the methyl ketone was injected into the glass wool. Each flush lasted 20 s. Each flush was quantitated to determine the amount of unbound methyl ketone. After flushing at 1650 s, 1 µl of the same methyl ketone was reinjected onto the glass wool at the top of the glass wool at the top of the sample tube, and the flushing cycle was repeated at 150 s intervials to 750 s. The samples were quantified for the amount of methyl ketone released for each purge to determine the diffeence in binding from the first injection. Five replications were made for each sample.

Data for µmoles of methyl ketones released after each of the two injections and the differences in concentrations eluted were compared over the entire flushing period using a Waller/Duncan comparison of means under SAS PROC GLM[29].

RESULTS and DISCUSSION

Perception of Flavor Compounds in the Presence of Milk Proteins

Figure 1 indicates vanillin flavor intensity relative to the 78.5 ppm vanillin reference standard for different concentrations of CAS and WPC. Figure 2 indicates the intensities of benzaldehyde, citral and d-limonene flavor intensity relative to the reference standards (benzaldehyde, 17.8 ppm; citral, 19.8 ppm; or d-limonene, 53.0 ppm) at differing concentrations of CAS and WPC.

The panelists noted greater flavor intensity of CAS and WPC as the protein concentrations increased. However, the protein flavor did not over power the note form the added flavor compound.

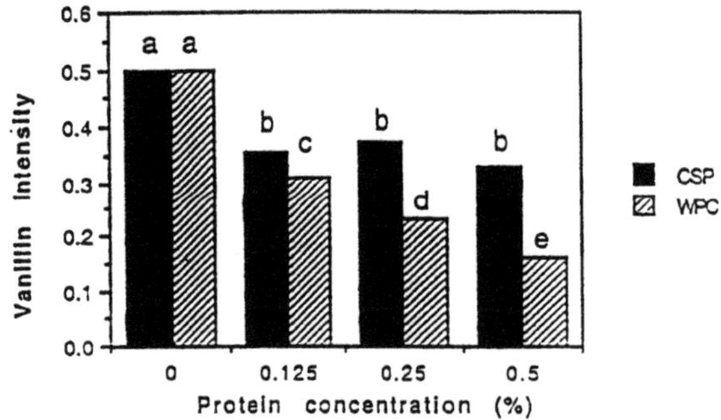

Figure 1. Vanillin flavor intensity relative to reference in the presence of sodium caseinate (CAS) and whey protein concentrate (WPC). Reference vanillin concentration was 78.5 ppm in a 2.5% sucrose solution. A sample with lower vanillin flavor intensity that the 78.5 ppm reference had a flavor below .5. Letters indicate significant differences between means (P < .05).

Vanillin flavor intensity (Fig. 1) decreased in the presence of CAS and WPC. Although there was no significant difference in vanillin flavor intensity (P < .05) with increasing CAS concentration, vanillin flavor declined from .32 (moderately less than reference) to .15 (much less than reference as WPC increased from .125 to 5%.

Benzaldehyde flavor intensity (Fig. 2a) significantly dropped (P < .05) from .45 (slightly less than reference) to .25 (much less than reference as the WPC concentration increased from 0 to .5%. There was no significant difference in benzaldehyde flavor concentration as the concentration of CAS increased.

Citral flavor intensity (Fig. 2b) showed no significant drop (P < .05) in intensity as the CAS concentration increased from 0 to .5%. Although citral flavor dropped from .41

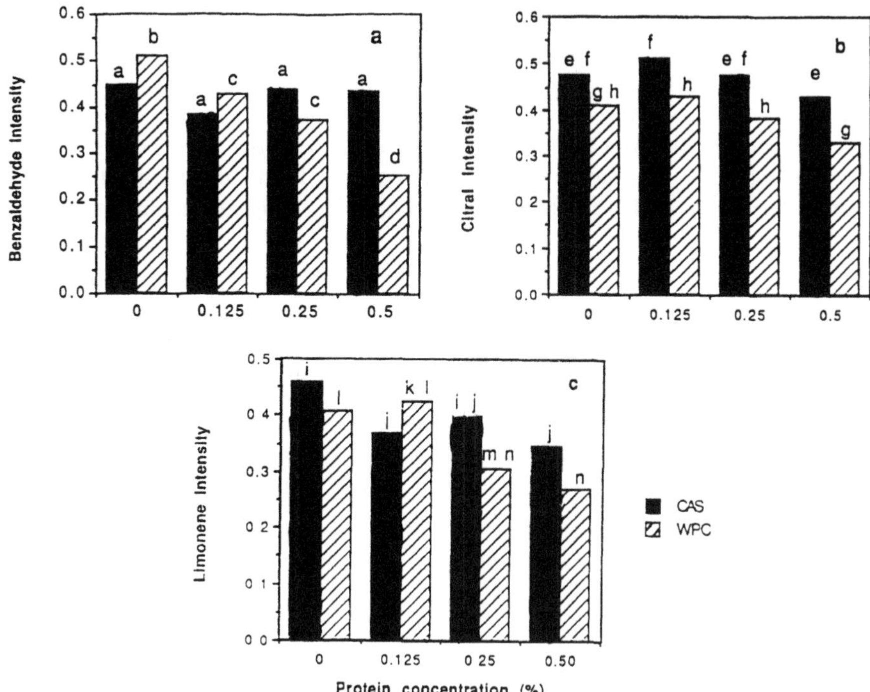

Figure 2. Flavor intensity relative to reference in the presence of increasing concentrations of sodium caseinate (CAS) and whey protein concentrate (WPC) for benzaldehyde (a), citral (b), and d-limonene (c). Reference concentrations are benzaldehyde (17.8 ppm), citral (19.8 ppm), and d-limonene (53 ppm) in a 2.5% sucrose solution. A sample with lower flavor intensity than the reference had a flavor intensity below .5. Letters indicate significant differences between means (P < .05).

(slightly less that reference) to .33 (less than reference as WPC increased from 0 to .5%, this decline was not significant.

d-Limonene flavor intensity (Fig. 2c) dropped significantly (P < .05) in the presence of CAS and WPC. The decline in flavor intensity was most marked for WPC, for which the intensity dropped from .41 (slightly less that reference to .27 (much less than reference) as WPC concentration increased from 0 to .5%.

Differences in flavor loss between CAS and WPC are expected because of differences in their structure and amino acid composition. The CAS and WPC were subjected to different levels of denaturation due to different processing conditions during manufacturing. Sodium caseinate forms a very loose, open micellar structure with hydrophillic and hydrophobic patches[32], which cause it to highly water soluble and to posses surfacantant properties[5]. Production of CAS causes minimal protein denaturation, even though their are acid precipitated at pH 4.6, washing, resolubilization using sodium hydroxide at pH 6 - 7, and spray drying[21].

Whey protein concentrate consists of a variety of proteins: β-lactoglobulin, α-lactalbumin, immunoglobulins, and bovine serum albumin[5]. These proteins are heat sensitive and more susceptible to denaturation during processing than casein[20]. Although BSA can interact with carbonyls[3], β-lactoglobulin is most likely to be involved in the flavor compound-WPC interaction because it binds aromatic compounds[7], serves as a transport protein for vitamin A[26,27], and constitutes 75% of the protein fraction in WPC[20].

Vanillin, benzaldehyde and *d*-limonene may interact with the retinol-binding site or with other sites near the surface of the protein[19] resulting in the decreased concentration the flavor compounds available for perception by the panelist. At physiological pH (6.8), retinol binding is most extensive[32], and the reactive thiol group of β-lactoglublin is exposed[7], which can react with aldehydes[31]. In contrast, a protective effect occurs for citral in citric acid solutions when casein is present[9]. Citral may form a more open structure in solution so there may be fewer strong interactions within nonpolar binding sites on β-lactoglobulin.

Binding of Methyl Ketones to Milk Protein Powder

Analysis of the delipidated milk protein on the Quatnachrome revealed a surface area of .873 m^2/g which was held constant throughout the experiment. In the two stage injection sequence, the difference in release of methyl ketone between the two injections indicated that binding may have occurred. This binding would be most apparent when

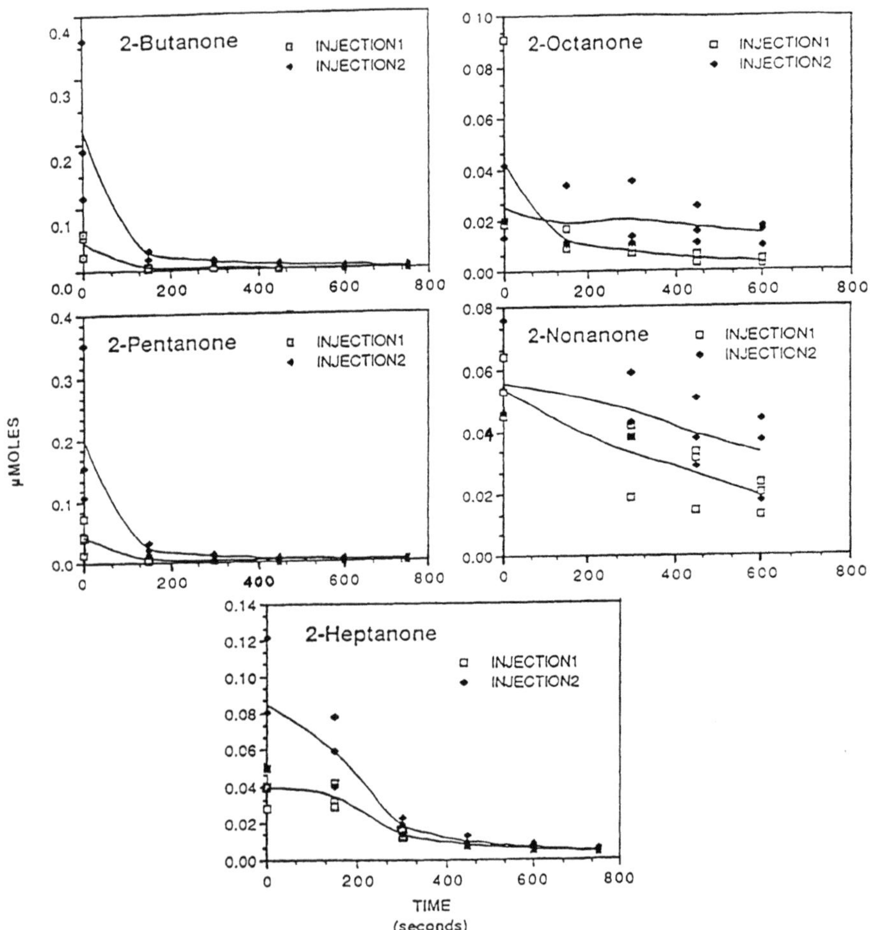

Figure 3. Plot of mean concentration of µmoles (five replicates) released for straight chain methyl ketones, 2-butanone, 2-pentanone, 2-heptanone, 2-octanone and 2-nonanone, versus time. One µL of ketone was individually injected onto different .1 g samples of delipidated milk protein powder.

there was low release of ketone on the first injection and high release on the second injection. In this process, the ketones interacted with the protein powder following injection 1. Subsequently, the ketone injected during injection 2 was unable to bind to the protein to the same extent as the ketone in injection1, and, therefore, the ketone in injection 2 was more rapidly removed during the flushing cycle. Because a larger percentage of ketone was unbound, more ketone was eluted in the first flush of injection 2.

Figures 3 and 4 show the mean concentrations (five replications) of ketones eluted during the flushing sequence of injections 1 and 2 of the straight and branched chain ketones. In the case of 2-butanone, .046 µmoles were released on the first flush of injection 1, and the amount declined with subsequent flushes. This pattern was also evident for injection 2 where .221 µmoles were released on the initial flush. After the third flush, most of the unbound ketones were removed during the flushing sequence for both injections 1 and 2. 2-Pentanone and 2-heptanone followed the same pattern of removal of unbound ketones within the first three flushes.

2-Octanone and 2-nonanone behaved differently, and removal of the unbound ketones continued throughout the flushing period. For 2-nonanone .054 µmoles were released on the first flush of injection 1 and .056 µmoles were released on the first flush of injection 2. The concentration of µmoles eluted in subsequent flushes declined. These results may have been due to molecular size and conformation of the ketones.

The branched chain methyl ketones behaved similar to straight chain methyl ketones where there was rapid removal of the ketone, 3,3-dimethyl-2-butanone, yielded a lower release with only .029 µmoles on the first flush of injection 1 and .036 µmoles on the first flush of injection 2. The 2,4-dimethyl-3-pentanone released .041 µmoles and .057 µmoles

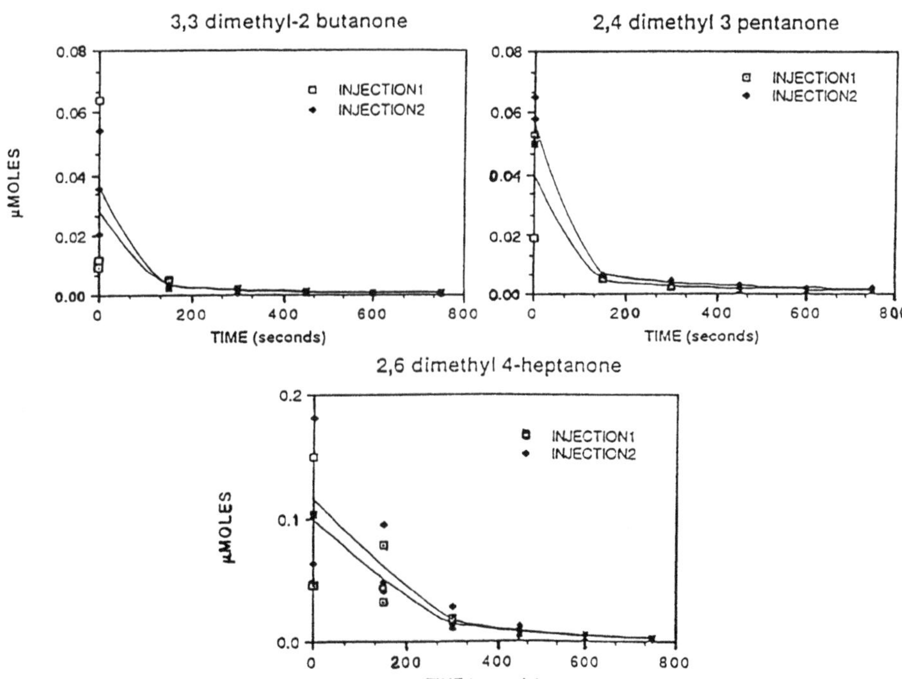

Figure 4. Plot of mean concentration of µmoles (five replicates) released for branched chain methyl ketones, 3,3-dimethyl-2-butanone, 2,4-dimethyl-3-pentanone, and 2,6-dimethyl-4-heptanone, versus time. One µL of ketone was individually injected onto different .1 g samples of delipidated milk protein powder.

on the first flush of injections 1 and 2, respectively. While 2,6-dimethyl-4-heptanone yielded .099 µmoles and .116 µmoles on injections 1 and 2 with the first flush, respectively. These results indicate that the greatest quantitative release occurred with the branched heptanone and the least with the branched butanone, although the difference was not significant (P = .07).

The concentration of µmoles released on the first flush of both injections 1 and 2 were significantly different (P = .07) from the later flushes. After combining the data for the first flush into straight and branched chain groups, significant differences (P = .07) were observed. The difference in concentration of ketones eluted between the two injections was related to binding. The higher binding characteristic of 2-butanone (difference = .175 µmoles) compared to 2-nonanone (difference = .002 µmoles) was probably due to molecular size and polarity. 2-Pentanone behaved much like 2-butanone and had a similar degree of binding. The degree of binding for 2-butanone and 2-pentanone was significantly greater (P = .07) than for 2-octanone and 2-nonanone. For the branched chain ketones, binding was highest for 2,6-dimethyl-4-heptanone (difference = .017 µmoles). However, this difference was not significantly greater than the difference for 3,3-dimethyl-2-butanone and 2,4-dimethyl-3-pentanone.

This research demonstrates that milk protein powder binds methyl ketones. Short chain methyl ketones exhibit a greater degree of binding than long chain methyl ketones. This observation may be a result of steric hindrance or inability of the larger methyl ketones to penetrate the milk protein. Because proteins influence the perceived flavor of foods by binding off-flavors[33], it is important that stored milk powder be protected from foreign flavor compounds and odors.

CONCLUSION

These studies have shown that the intensity of vanillin, benzaldehyde, and d-limonene flavor decreased as the concentratin of WPC increased. However, an increased concentration of CAS not reduce the perception of these flavor compounds. Citral, a more open structure in solution, did not exhibit an decrease in flavor perception in the presence of CAS or WPC. Additonally, this research demonstrated that delipidated milk protein powder binds methyl ketones. Short chain methyl ketones exhibit a greater degree of binding than long chain methyl ketones. the ability of milk proteins to bind flavor compounds may result in the lowering of desired flavor compounds that are added to a dairy product or result in the unitentional additon of undesirable flavor compounds into food products manufactured from milk protein powder.

REFERENCES

1. E. Berlin, B.A. Anderson, and M.J. Pallansch, Influence of dehydration method on the adsorption of benzene vapor by dried casein, *J. Colloid Interface Sci.* 43:571 (1973).
2. J.E. Cometto-Muniz, Odor, taste and flavor perception of some flavoring agents, *Chem. Senses* 6:215 (1981).
3. S. Damodaran and J.E. Kinsella, Flavor-protein interactions: Binding of carbonyls to bovine serum albumin: thermodynamic and conformational effects, *J. Agric. Food Chem.* 29:1253 (1980).

4. E. Defour and T. Haertle, Binding affinities of α-ionone and related flavor compounds to β-lactoglobulin: Effects of chemical modifications. *J. Agric. Food Chem.* 38:1691 (1990).

5. G. Doxastakis, G.. Milk proteins, in: *Food Emulsifiers, Chemistry, Technology, Functional Properties and Applications.* Charalambous, G., and Doxastakis, G., eds. Elsevier Sci. Publ., Amsterdam, Neth. (1989).

6. R.G. Einig, "Interactions of meat aroma volatiles with soy proteins." Doctoral Dissertation, University of Missouri, Columbia (1983).

7. H.M. Farrell, Jr., M.J. Behe, and J.A. Enyeart, Binding of *p*-nitrophenyl phosphate and other aromatic compounds by β-lactoglobulin, *J. Dairy Sci.* 70:252 (1987).

8. K.L. Franzen and J.E. Kinsella, Parameters affecting the binding of volatile flavor compounds in model food systems: I. Proteins, *J. Agric. Food Chem.* 22:675 (1974).

9. H. Friedrich and B.A. Gubler, Wechselwirkingen zwischen Aromstoffen und Lebensmittel: II. Die instabilitat von citral in zitronensaft, *Lebensm Wiss. Technol.* 11:215 (1978).

10. H. Gremli, Interaction of flavor compounds with soy protein, *J. Am. Oil Chem. Soc.* 51:95A (1974).

11. A.P. Hansen, Z. Haque, and J.E. Kinsella, Parameters affecting the binding of volatile flavor compounds in model food systems: I. Proteins, *J. Agr. Food Chem.* 22:675 (1974).

12. A.P. Hansen and J.J. Heinis, Decrease of vanillin flavor perception in the presence of casein and whey proteins. J. Dairy Sci. 74:2936 (1991).

13. A.P. Hansen and J.J. Heinis, Benzaldehyde, citral, and *d*-limonene flavor perception in the presence of casein and whey proteins, *J. Dairy Sci.* 75:1211 (1992).

14. Z. Haque and J.E. Kinsella, Interactions between heated κ-casein and β-lactoglobulin: Predominance of hydrophobic interaction in the initial stages of complex formation, *J. Dairy Research* 55:67 (1988).

15. E. Jasinski and A. Kilara, Flavor binding by whey proteins, *Milchwissenschaft* 40:596-599 (1985).

16. G. Maier, Zur bindung fluchtiger aromastoffe an lebensmittel: VII. Aliphatische aldehyde, *Z. Lebensm. Unters. Forsch.* 151:384 (1973).

17. S.L. McMullin, R.A. Bernhard, and T.A. Nickerson, T.A.. Heats of adsorption of small molecules on lactose, *J. Agric. Food Chem.* 23:452 (1975).

18. O.E. Mills and J. Solms, Interaction of selected flavor compounds with whey proteins, *Lebensm. Wiss. Technol.* 17:331 (1984).

19. H.L. Monaco, G. Zanotti, P. Spadon, M. Bolognesi, L. Sawyer, and E.E. Eliopoulous, Crystal structure of the trigonal form of bovine β-lactoglobulin and of its complex with retinol at 2.4 A resolution, *J. Mol. Biol.* 197:695 (1987).

20. C.V. Morr and E.A. Foegeding, Composition and functionality of commercial whey and milk protein concentrates and isolates: A status report, *Food Technol.* 44(4):100 (1990).

21. L.L. Muller, Manufacture of casein, caseinates and co-precipitates, in: Developments in Dairy Chemistry 1. Fox, P.F., ed. Appl. Sci, Barking, Engl. (1982).

22. W.W. Nawar, Some variables affecting composition of headspace aroma, *J. Agr. Food Chem.* 19:1057 (1971).

23. R.C. Oldfield, Perception in the mouth. *Soc. Chem. Ind.* (Lond.) Monogr. 7:3 (1960).

24. T. O'Neill and J.E. Kinsella, Binding of alkonone flavors to β-lactoglobulin: Effects of conformational and chemical modification, *J. Agri. Food Chem.* 35:770 (1987).

25. T. O'Neill and J.E. Kinsella, Effect of heat treatment on modification on conformation and flavor binding by β-lactoglobulin, *J. Food Sci.* 53:9056 (1988).

26. M.Z. Papiz, L. Sawyer, E.E. Eliopoulos, A.C.T. North, J.B.C. Findlay, R. Sivprasadarao, T.A. Jones, M.E. Newcomer, and P.J. Kraulis, The structure of β-lactoglobulin and its similarity to plasma retinol-binding protein, *Nature* (Lond.) 324:383 (1986).

27. S. Pervais and K. Brew Homology of β-lactoglobulin, serum retinol binding protein, and protein HC, *Science* 228:335 (1985).

28. K.A. Robillard and A.Washnia, Aromatic hydrophobes and β-lactoglobulin: A. Thermodynamics of binding, *Biochem.* 11:3835 (1972).

29. SAS Institute, *SAS Users Guide: Statistics*, SAS Institute, Inc. Cary, NC (1982).

30. SAS Institute, *SAS Users Guide: Statistics*, Version 5 Edition.. SAS Inst., Inc. Cary, NC (1985).

31. M. Schubert, Compounds of thiol acids with aldehydes, *J. Biol. Chem.* 114:351 (1936).

32. H. Swaisgood, Structural changes in milk proteins, in: Milk Proteins. Barth, C.A., and Schlimme, E. ed. Steinkopff Verlag, Darmstadt, Germany (1989).

33. W. van Osnabrugge, How to flavor baked goods and snacks effectively, *Food Technology.* 43:74 (1989).

34. A. Washnia and T.W. Pender, Hydrophobic interactions in proteins: The alkane binding site of β-lactoglobulins A and B, *Biochem.* 5:1534 (1966).

35. S. Yoshida, Ye-Xiuyun, and T. Nishiumi, The ability of α-lactalbumin and β-lactoglobulin to bind mutagenic heterocyclic amines, *J. Dairy Sci.* 74:3741 (1991).

PRODUCTION OF WHEY-PROTEIN-ENRICHED PRODUCTS

Daniel M. Mulvihill and M.B. Grufferty

Department of Food Chemistry
University College
Cork, Ireland

INTRODUCTION

Whey is the serum liquid portion remaining after removal of fat and casein from milk during the manufacture of cheeses and caseins. Traditionally, whey was regarded in the dairy industry as an undesirable by-product and it was either fed in limited quantities to farm animals or treated as effluent, disposed of as a pollutant into streams and rivers, or irrigated onto pastures. However, environmental awareness led to the imposition of strict controls on the disposal of waste whey which necessitated the construction and operation of costly effluent treatment plants for whey treatment prior to disposal. This, together with a recognition that whey was a potentially valuable source of nutrients, stimulated interest in the development of commercially viable processes to convert liquid whey into viable animal and human food products.

Whey is now regarded as a co-product of cheese and casein manufacture and the objectives of the industry are to apply the best technologies available which allow the profitable processing of large volumes of high quality whey into products that can be advantageously used in a broad range of food, feed and industrial applications.

WHEY COMPOSITION

There are two principal types of whey, sweet whey (minimum pH 5.6) from the coagulation of milk by rennet-type enzymes in the manufacture of several cheeses and rennet casein and acid whey (maximum pH 5.1) from the coagulation of milk primarily with acid in the manufacture of acid cheeses and acid casein. The composition of whey varies depending on milk composition, cheese variety, casein type and processing conditions used in the manufacture of the cheese or casein. However, whey typically contains ~ 50% of the total solids of the original milk. Both acid and rennet whey types contain 0.65-0.8% (w/v) protein which corresponds to 10-12% (w/w) protein on a dry basis. The chief constituent is lactose, which represents approximately 70% of the solids. Acid wheys produced as a result of fermentation have a lower lactose content than mineral acid whey, as some of the lactose is metabolised to lactic acid by the fermenting bacteria. Acid whey contains a higher concentration of minerals than rennet whey due to dissolution of the colloidal calcium phosphate component of the casein micelles during acidification. Cheese whey derived from whole milk has higher concentrations of residual milk fat than that derived from skim milk. The different mineral compositions and pH values are important factors in whey processing and may have significant effects on the functionality of dried whey and whey protein-enriched products.

Pre-processing

To obtain products of acceptable uniformity and storage stability certain pretreatments of the 'raw' whey are necessary.

Clarification. When the whey is recovered from a cheese or casein manufacturing operation it is inevitable that low levels of 'curd' fines are present. These confer a serious risk of blocking heat exchanger channels or damaging ultrafiltration (UF) or reverse osmosis (RO) membranes. In addition, they may adversely affect the solubility properties and flavour of the end products.

Clarification is therefore necessary and is usually achieved by either a combination of settling, screening and decanter centrifugation or by decanter centrifugation alone depending on the size and level of fines.

Separation. When whey is recovered from cheese made from whole milk it usually has a substantial fat content. For economic reasons, and for flavour stability of whey products, it is normal to remove the fat from the whey. In practice this is usually achieved using a self-discharging separator. In a low fines load situation the separator may also serve to clarify the whey or act as a second-stage-clarifier following the primary clarifier in a high-fines load situation.

Despite the close to total removal of 'free' fat from whey centrifugally, it is not possible to remove all the fat by this means. A residual fat level of about 0.02% is inevitable in whey from cheese manufactured from whole milk unless additional processes are utilized.

Pasteurization. It is essential to pasteurize cheese whey immediately after curd-whey separation if optimum microbiological quality and storage stability is to be achieved. If the whey has to be stored before heat treatment it must be cooled to 5°C or below as quickly as possible. The time-temperature combination for pasteurization should be in the range 72-75°C for 15-20 s to ensure proper reduction in viable counts and to inactivate phosphatase and chymosin enzymes, but may in practice be up to 80°C to minimize the risk of bacteriophage contamination of the cheese factory.

During subsequent processing, whey solutions are usually pre-heated and thermally concentrated at optimum temperatures and holding times and maintained at low temperatures during storage to minimize microbial growth and to optimize/minimize physico-chemical modification of the proteins and other whey constituents.

Whole Whey Powders

The simplest whey processing operation is when whey is dehydrated *per se* to produce whole whey powder with a protein content of 10-12% (w/w). After the pre-processing stages the whey is usually concentrated to 40-60% total solids by evaporation under vacuum in the multiple stage evaporator at temperatures below 70°C to avoid protein denaturation. Reverse osmosis may be employed to concentrate the whey to approximately 20% total solids to increase the capacity of the evaporator. The next step in production of dried powder from whey concentrate should be a final dehydration process such as spray drying; however, the process is not that simple and direct drying of fresh concentrate is not desirable.

The structural elements of a directly dried whey particle are lactose, either amorphous or partially crystalline, whey protein particles, fat in globular or non-globular (free fat) forms, and air as spherical cells. The fat, protein and air are dispersed in a continuous phase of amorphous lactose. The distribution of lactose between the amorphous glass and crystalline states depends on the processing conditions. Most of the lactose in directly dried whey powders is in the amorphous or glass state with a ß-lactose to α-lactose ratio of 1.25 to 1.6. Owing to the high hygroscopicity of lactose glass, the dried particles readily adsorb moisture from the surrounding atmosphere. When this happens, the concentration of lactose is diluted so that its molecules acquire sufficient mobility and space to orientate themselves into a crystal lattice which leads to clumping and 'caking' or compaction of powder particles to a hard mass. Lactose crystallization leads to protein insolubility, de-emulsification of fat and accelerated flavour deterioration. Another consideration is that the solids level of the concentrate is limited in direct drying processes and this results in poor process economy.

Crystallization. Special methods have been developed to produce precrystallized, non-hygroscopic, non-caking and wettable whey powders. Following concentration to 50-60% total solids, the concentrate is flash cooled to about 30°C, pumped to crystallization tanks fitted with cooling jackets and agitators, seeded with either finely ground α-lactose monohydrate or well-crystallized whey powder, held for 3-4 hours, then further cooled slowly at a rate of about 3°C/h to as low as 10°C to crystallize up to 80% of the lactose as α-lactose monohydrate which is less hygroscopic than anhydrous ß-lactose. Alternatively, the flash cooled concentrate may be pumped to an ordinary non-jacketed agitation tank corresponding in volume to 2-3 hours production and the concentrate seeded as described above. When the tank is full the product is discharged at the same speed as fresh product is fed into it (backmixing). The discharged product is flash cooled to 15°C while being pumped into a secondary crystallization tank which has the same volume as the primary tank. When this second backmix-tank is full the crystallized concentrate is fed to the drier. This latter process is a continuous short-time crystallization process.

Drying. Irrespective of the drying technology, pre-crystallization of whey concentrate is a common process step. The crystallized concentrate may be spray dried in a single stage process to produce a dense pre-crystallized whey powder which has poor dissolving properties or, now more commonly, dried in a multi-stage process to produce more functional powders. In two stage drying, the pre-crystallized whey concentrate is first dried at low temperatures in a spray drying stage to give a product containing 5-8% moisture and post-crystallization and final drying take place in an external vibrating fluidized bed. One three stage drying process or spray drying with integral fluid bed, is an extension of the two stage concept but transfers the second drying stage into the spray drying chamber and has a final drying conducted in the third stage located outside the drying chamber. Another three stage drying process involves atomization in a low profile drying chamber, primary drying during fall of the atomized droplets towards an integrated conveyor belt. Post-crystallization and second and third stage drying and cooling take place on the belt as it moves at the base of the drier. The whey powder products produced in these 2 and 3 stage processes have large agglomerates, low bulk density, 85-95% crystallized lactose and are non-hygroscopic, non-caking and readily rehydratable.

The process described above involve dehydration and recovery of essentially all the solid constituents of the whey in one product and thus the product has a low protein content; however, processes have been developed which allow fractionation of the solid constituents and production of a large range of protein-enriched whey products.

Demineralized Whey Products

The high ash content of whole whey can adversely affect the flavour and nutritional quality of whey products. Thus, processes are used to demineralise whey. Demineralization is particularly useful when applied to sweet whey; powders with a very low mineral content, suitable for use in baby and special diet foods can be produced. Also, it is well recognised that yield and purity in lactose manufacturing operations are enhanced when the mineral content of the lactose feedstock is reduced.

Electrodialysis, ion exchange and nanofiltration or 'loose' reverse osmosis are utilized for the demineralization of whey. Electrodialysis has been shown to be cost-effective for up to 70% mineral removal from cheese whey. Ion exchange chromatography is suitable for removal of up to 90% of the minerals in whey. The commercial literature generally claims that about 50% demineralization is economically feasible with nanofiltration or 'loose' reverse osmosis; an added advantage of the process is that water is also removed and thus a significant degree of concentration (up to 24% total solids) is also achieved. Many modern whey demineralization plants use a combination of processes and the mineral profile of the whey products can now be carefully controlled.

Lactose and Delactosed Whey Products

Lactose and reduced lactose whey products are produced by concentration of whey, lactose crystallization and recovery by centrifugation, washing and drying, and dehydration of the delactosed (mother) liquor. Typically, whey is concentrated by evaporation to 58-62% total solids at temperatures below 70%, the concentrate is slowly cooled and seeded to induce lactose crystallization and the mother liquor (delactosed whey) is separated from the lactose crystals by decanting. The crystals are usually washed in the decanter to remove impurities (minerals) and may be dissolved and recrystallized prior to drying to yield a product of high purity; the mother liquor may be concentrated and dried to produce delactosed whey powder (or WPC) containing about 25% protein on a dry basis or may be further processed by UF to produce WPC's with a higher protein content.

Delactosed whey powder has a high mineral content (up to 25%) which restricts its use in some food applications. Therefore, demineralization is often used in conjunction with lactose removal. Added advantages are less scaling in the evaporator due to insoluble calcium phosphate and calcium citrate salts, less impurities in the recovered lactose and production of demineralized-delactosed whey powder (reduced minerals WPC). In a typical process for the recovery of high quality lactose and demineralized-delactosed whey powder, pasteurized whey or whey concentrated to 20-30% total solids is adjusted to pH 6.2-6.4 and demineralized by electrodialysis or a combination of electrodialysis and ion exchange processes. The demineralized whey is concentrated to 40-60% total solids by evaporation at temperatures less than 70°C (to avoid protein denaturation). The concentrate is cooled in a controlled manner to induce lactose crystallization. The lactose is recovered by decanting and further processed and the mother liquor is spray dried to produce a powder containing 25-35% protein on a dry basis.

Alternatively, whey can be delactosed before demineralization. In this process whey is concentrated by evaporation to 50-60% total solids, cooled to induce lactose crystallization 40-60% of the lactose is removed by centrifugation. Residual lactose crystals in the mother liquor are dissolved by heating to 43-50°C. The mother liquor is clarified to remove insoluble protein, demineralized by electrodialysis and/or ion exchange processes, further concentrated and spray dried.

Whey Protein Concentrate (WPC) and Whey Protein Isolate (WPI)

Fractionation of whey to recovery whey protein concentrates (WPC) and whey protein isolates (WPI), whey protein powders containing 35-95% protein on a dry basis, is accomplished on a commercial scale by ultrafiltration-diafiltration and ion exchange chromatography.

Ultrafiltration-Diafiltration. Ultrafiltration (UF) is a pressure membrane filtration process that facilitates the selective separation of whey proteins from lactose, salts and water under mild conditions of temperature and pH. It is a physico-chemical separation technique in which a pressurized whey solution flows over a porous membrane that allows the passage of only relatively small molecules. The retained solution (retentate) flows over the membrane, while under the influence of pressure, water flows through the membrane, together with low molecular weight solutes (the permeate). The protein is retained by the membrane and is therefore concentrated relative to other solutes in the retentate. Fat globules, suspended solids and bacteria are also retained.

The membranes used in UF are asymmetric microporous structures, the effective layers of which appear to contain pores with diameters ranging from 1 to 20 nm. Commonly used membrane configurations include tubular, plate and frame, spiral-wound and hollow fibre, with each configuration offering advantages and disadvantages for particular applications. The membranes used are ceramic or synthetic polymer (e.g. polysulphone or polyamide) membranes which have high resistance to high temperature (up to 100°C), can withstand a wide pH range (1-13) and can be cleaned with agents normally used in the dairy industry (e.g. HNO_3 and NaOH). Although UF is currently the method of choice for commercial manufacture of WPC of varying protein concentration, it has several major problems that limit its operational performance. These problems include: high capital and operating cost; membrane fouling with concomitant loss of permeate flux rate; incomplete removal of low molecular weight solutes unless diafiltration (dilution of retentate with water and repeated UF) is used; cleaning, sanitation and related microbial problems; production of large volumes of permeate for further processing or disposal.

Prior to UF processing, whey is commonly pre-treated by methods involving pH and/ or temperature adjustments, addition of calcium or calcium complexing agents and either quiescent standing, centrifugation or microfiltration to dissolve colloidal calcium phosphate and/or to remove insoluble cheese curd or casein fines, milk fat and calcium lipophosphoprotein complexes (Hayes et al., 1974; Breslau et al., 1975; de Wit and de Boer, 1975; Lee and Merson, 1976; de Wit et al., 1978; Matthews et al., 1978; Muller and Harper, 1979; Maubois et al., 1987). These pretreatments increase flux during ultrafiltration, prevent fouling of the membrane and modify the properties of the whey concentrates.

Because of the variation in feed rate throughout the time of UF (initially the flux is high, falling quickly due to concentration polarization and then more slowly as the membrane fouls), it is usually necessary to build up a buffer stock of whey before UF commences. The UF process is generally operated at temperatures above 50°C to minimize microbial growth. Most modern UF plants are continuous multiple-stage in series recirculation systems. Each stage consists of a number of individual membrane assemblies, or modules, a recycle loop to return some of the retentate back to the feed of the stage and associated pumps and instrumentation. During operation a high volumetric flowrate is maintained within each membrane assembly by recirculating a large proportion of the retentate back to the feed of the stage. Flow through the plant is most commonly controlled by continuous ·withdrawal of retentate, at the desired concentration, from the last stage. Steady-state operation is maintained by the introduction of sufficient feed into the first stage to compensate for the volume of permeate removed from all stages and the volume of retentate withdrawn. The use of a continuous multi-stage operation allows large volumes of feed to be concentrated with a short average residence time in the ultrafiltration plant. It also requires considerably less pumping energy than a batch mode of operation because the whole of the operating pressure is not lost on each recycle. There is a maximum ratio of protein to total solids in the retentate which can be achieved by UF on modern plants. Above this ratio the viscosity of the retentate is such that the flux rate becomes excessively low. Where it is necessary to increase the ratio beyond that achievable by ultrafiltration alone, diafiltration is commonly used. In diafiltration, water is added to the retentate prior to further ultrafiltration, as a means of increasing the removal of membrane-permeable species. This is achieved by dilution of the retentate either before or during additional ultrafiltration. The latter case is most common in continuous multi-stage operation, where ideally a constant retentate volumetric flowrate is maintained and diafiltration water is added to the feed of a number of individual stages at the same rate that permeate is being removed from each stage. With diafiltration, ratios of protein to total solids up to $\sim$ 0.80:1.0 in the final retentate can be achieved.

The retentate from the ultrafiltration plant is usually cooled to $\sim$ 4°C and stored until a sufficient volume has been accumulated for drying. Pasteurization of the retentate using a heat treatment of 66-72°C for 15 s may also be necessary to reduce the number of bacteria, because bacteria in the whey are concentrated during ultrafiltration. Food grade chemicals may be added to the retentate to modify the physico-chemical properties of the WPC.

In most commercial plants the final retentate is relatively dilute and some further concentration is required before spray drying. Low temperature evaporation ($\sim$ 50°C) to 25-44% (w/v) total solids or 15-20% (w/v) protein in the concentrate is the normal commercial practice. Drying results in the production of powders containing 35-80% (w/w) protein.

Ion Exchange Adsorption. Whey proteins are amphoretic molecules and therefore have a net charge that depends on solution pH. At pH values lower than their isoelectric point (pH $\sim$ 4.6 for major whey proteins), whey proteins have a net positive charge and behave as cations which can be adsorbed on cation exchangers. At pH values above their isoelectric point, the proteins have a net negative charge and behave as anions which can be adsorbed on anion exchangers. Media with suitable pore sizes and surface characteristics have been developed specifically for the recovery of proteins from dilute solutions, depending upon the pH of the medium. Two major ion exchange fractionation processes have been commercialized for the manufacture of WPI.

The "Vistec" process uses a cellulose-based exchanger in a stirred tank reactor (Burgess and Kelly, 1979; Palmer, 1982). The process involves a series of steps that are performed as a fractionation cycle: (1) whey is adjusted to pH < 4.6 with acid, pumped into a tank reactor and stirred to allow protein adsorption onto the ion exchanger, (2) lactose and other unadsorbed materials are filtered off, (3) the resin is resuspended in water and the pH adjusted to > 5.5 with alkali to release the proteins from the ion exchanger, (4) the aqueous solution of proteins is separated from the resin by filtration in the tank reactor, is concentrated by ultrafiltration and evaporation and is spray dried to give WPI containing $\sim$ 95% protein. UF treatment of the protein-rich eluate fraction is essential for purification and concentration of the protein.

The "Spherosil" processes (Mirabel, 1978; Kaczmarek, 1980) use either cationic Spherosil S or anionic Spherosil QMA ion exchangers and fractionation is accomplished in fixed-bed column reactors. Acidified whey at pH < 4.6 is applied to the Spherosil

S column reactor to allow protein adsorption by the strongly acidic cation resin. After lactose and other unadsorbed solutes have been eluted, the pH is raised by addition of alkali to elute absorbed proteins from the reactor. The protein-rich eluate fraction is concentrated by UF and evaporation and spray dried to produce WPI. Sweet whey at pH > 5.5 is applied to the Spherosil QMA column reactor to permit negatively charged protein molecules to be adsorbed by the strong anionic ion exchanger. After elution of non-protein materials, the proteins are released by lowering the pH with acid. Released proteins are concentrated and spray dried as for the Spherosil S process.

These adsorption processes recover ~ 85% of the protein under ideal operating conditions and the recovered concentrates are characterized by high protein and low lactose, lipid and ash concentrations and have good functionality.

Lactalbumin

Whey proteins are globular and are readily denatured on heating. On transformation from their globular conformations to more random structures, sulphydryl and hydrophobic groups are exposed and protein-protein interactions occur. The extent of aggregation and precipitation of the denatured proteins depends on heating temperature and holding time, pH and concentration of calcium. Commercial precipitation conditions employed to recover heat denatured protein depend on whey type and the desired final product characteristics and whey may be preconcentrated and/or demineralized prior to precipitation. The precipitated protein, referred to as lactalbumin, may be recovered by settling and decanting, vacuum filtration, self-desludging centrifuges or horizontal solid-bowl decanters. The precipitate may be washed to reduce mineral and lactose contents and dried in spray, roller, ring or fluidized bed driers. Protein yields may be up to 90% of that in the whey and lactalbumin containing up to 90% protein on a dry basis may be recovered, depending on precipitation pH and degree of washing.

Electrochemical Coagulation of Whey Protein

A laboratory scale electrochemical coagulation (EC) process was recently described by Janson and Lewis (1994) for the recovery of protein from cheese whey. The process used an electrolytic cell design in which the cathode and anode compartments were separated by a non ion-selective membrane (fabricated from cotton).

EC is a process based on electrolysis of water which results from two processes namely an electrode/solution boundary process and a diffusion/convection process, which occur simultaneously. The electrode/solution boundary process is where transfer of electric charge occurs at the surface of the electrode. At the cathode the predominant electrochemical reaction (ECR) in acid conditions is $2H^+ + 2e = H_2$ and the predominant ECR in alkaline conditions is $2H_2O + 2e = 2OH^- + H_2$. At the anode the predominant ECR in acid conditions is $2H_2O = 4H^+ + O_2 + 4e$ and the predominant ECR in alkaline conditions is $4OH^- = 2H_2O + O_2 + 4e$. All these reactions take place at the surface of the electrode and dramatically change the acid-base conditions in the thin boundary layer (0.1-0.5 mm). Diffusion and convective processes are also important, as these are responsible for transferring the electronic charge to the bulk of the electrolyte solution.

An increase in acidity in the anode compartment and alkalinity in the cathode compartment takes place due to the membrane reducing ion transport between electrode compartments. In the presence of the membrane there is an increase in acidity or alkalinity in the bulk of the solution; without the membrane it only occurs in the boundary layers.

The most effective process for the recovery of protein from whey was where whey (pH 6.9) first moved through the anode compartment in a thin channel (3 mm wide). The flow velocity (0.9-1.0 ml min^{-1}) and voltage (30 V) were optimized to give pH values appropriate for coagulation in the anode compartment. After emerging from the anode compartment the whey was separated into a coagulum and a protein depleted whey and the latter was pumped through the cathode compartment to complete the circuit. The optimum pH of the whey emerging from the anode compartment was 4.6 which resulted in a 73.5% reduction in protein content between the whey (0.71% w/w protein) and the protein depleted whey (0.19% w/w protein) and an exit pH of 7.43 for the protein depleted whey emerging from the cathode compartment. Chemical coagulation by adjustment to pH 4.6 using HCl only resulted in an 8% reduction in protein content between the whey and chemically protein depleted whey.

A conclusive mechanism for coagulation in the EC process is not known; however, it has been proposed that the strong hydration layer which maintains the colloidal stability of whey proteins is very much reduced by the interaction of pH reduction and the applied external electric field or alternatively the external electric field could markedly influence ion-oriented motion and accelerate charge transport through the hydration layer, resulting in an easier loss or charge from the protein.

Fractionation of Whey Proteins

The proteins in whey include ß-lactoglobulin (ß-lg), α-lactalbumin (α-la), bovine serum albumin (BSA), a number of biologically active proteins, including immunoglobulins (Igs), lactoperoxidase and lactotransferrin, together with a number of minor proteins, including protease-peptones and κ-casein glycomacropeptide which is present in whey produced as a result of enzymatic coagulation of casein.

Methods for the isolation of individual whey proteins on a laboratory scale by salting-out, ion exchange chromatography and/or crystallization have been available for about 40 years. Owing to the unique functional, physiological or other biological properties of many of the whey proteins, there is an economic incentive for their isolation on an industrial scale. For example, ß-lg, the principal whey protein in bovine milk, produces better thermo-set gels than α-la (de Wit et al., 1988). However, human milk does not contain ß-lg which is the most allergenic of the bovine milk proteins for the human infant; therefore, α-la would appear to be a more appropriate protein for the preparation of humanized baby formula than total whey protein.

ß-Lactoglobulin and α-Lactalbumin. A number of methods have been developed for the separation of α-la and ß-lg. The most commercially amenable of these are broadly based on the method developed by Pearse (1983) in which the low heat stability of apo-α-la is exploited to precipitate it from whey, leaving ß-lg, BSA and immunoglobulins (Igs) in solution. α-La is a Ca-containing metalloprotein which is denatured at relatively low temperatures but renatures on cooling. The protein loses its Ca and is transformed into the apo form on acidification to $\leq$ pH 5. In the apo form it aggregates on heating to $\sim$ 55°C and can be fractionated from the soluble proteins by centrifugation, filtration or microfiltration. This method has been modified by Pearse (1987).

A basically similar method was described by Pierre and Fauquant (1986) who used clarified UF concentrate as starting material. It was claimed that clarification improved the fractionation; thus, the ß-lg fraction was 98% pure but the α-la fraction was contaminated with BSA and unidentified proteins.

A further modification of the process was proposed by de Wit and Bronts (1994). It comprises the following steps: (i) incubating a whey protein solution with a calcium-binding ion exchange resin in its acid form to initiate the destabilization of the α-la, (ii) adjusting the treated protein solution to a pH value of 4.3-4.8, after separation of the resin, (iii) incubating the protein solution at a temperature of 10-50°C to promote the flocculation of α-la, (iv) fractionating the proteins in the solution at pH 4.3-4.8, to give an α-la-enriched fraction and a ß-lg-enriched fraction, (v) raising the pH of the α-la-enriched fraction sufficiently to solubilize it, (vi) raising the pH of the ß-lg-enriched fraction sufficiently to neutralize it, (vii) downstream processing and drying of both fractions.

Stack et al. (1995) described a very integrated process for the recovery of α-la and ß-lg-enriched fractions from raw whey. The process involved: (i) demineralization of raw whey by a combination of electrodialysis and ion-exchange to reduce the calcium content to below 120 ppm, on a dry weight basis and the pH to < 3.4, (ii) heating to 71-98°C for 50-95 sec followed by rapid cooling to ~ 10°C, (iii) two-stage concentration to 55-63% total solids at < 69°C with optional further demineralization between stages, (iv) cooling to induce lactose crystallization, (v) separation of lactose crystals from whey protein liquor, (vi) adjusting the whey protein liquor to pH 4.3-4.7 at < 10°C and then heating to 35-54°C for 1-3 h to induce flocculation of α-la, (vii) separating the flocculant α-la-enriched fraction from the ß-lg-enriched supernatant by mechanical or membrane separation, (viii) washing the α-la-enriched flocculant with a solution isoionic with the whey protein liquor and with a pH of 4.3-4.7, (ix) neutralization, downstream processing and spray drying of both fractions.

α-La and ß-lg are insoluble in pure water at their isoelectric points; ß-lg requires a higher ionic strength for solubility than α-la, a characteristic which has been exploited by Amundson et al. (1982) and Slack et al. (1986) to fractionate α-la and ß-lg. Whey is concentrated by UF, acidified to pH 4.65 and demineralized by electrodialysis to < 0.023% ash, ß-lg precipitates and may be recovered by centrifugation with a yield of > 90%.

A method for the removal of ß-lg from whey using $FeCl_3$ was described by Kuwata et al. (1985). On adjustment of whey containing 4 mM $FeCl_3$ to pH 3.0, α-la and Ig precipitate, leaving ß-lg in solution. Fe^{3+} can be removed from the α-la fraction by ion exchange or ultrafiltration. An alternative chemical fractionation procedure was reported by Al-Mashikhi and Nakai (1987a) using sodium hexametaphosphate (SHMP). Optimum conditions were 1.33 g SHMP/l at pH 4.07; after holding at 22°C for 1 h, > 80% of the ß-lg had precipitated, leaving most of the α-la and Ig in the supernatant. This method was modified by Cuddigan (1991) who used UF retentate and *pro rata* increases in SHMP; unfortunately, different degrees of concentration were required to obtain the purest preparations of ß-lg in the precipitate and of α-la in the supernatant.

ß-Lg and α-la/Ig may also be prepared from UF retentate (volume concentration factor of 10) by fractionation with 7% (w/v) NaCl at pH 2.0 (Mailliart and Ribadeau-Dumas, 1988). Under these conditions, ß-lg remains soluble while all other proteins precipitate. Essentially pure ß-lg can be precipitated from the supernatant at 30% (w/v) NaCl at pH 2.0, while the precipitate can be rendered essentially free of ß-lg by washing with 6% NaCl at pH 2.0.

A long established chemical method for the isolation of ß-lg is that of Fox et al. (1967) who used 3% trichloroacetic acid (TCA) to precipitate all whey proteins except ß-lg, a highly purified preparation of which can be obtained from the supernatant by salting-out or other suitable methods. In its original form, the method is probably not economical on a commercial scale but fractionation of a UF retentate may be economical; in any case, the use of TCA in the preparation of food-grade proteins may not be acceptable.

The ion exchangers used to recover WPI may also be used to fractionate whey proteins (Skudder, 1985). All the whey proteins are adsorbed initially on Spherosil

QMA but on continued passage of whey through the column, ß-lg, which has a higher affinity for this resin than the other proteins, displaces α-la and BSA giving a mixture of these proteins in the eluate; a highly purified ß-lg can be obtained by eluting the protein-saturated column with 0.1 M HCl.

A number of methods have also been described for the production of ß-lg-depleted whey protein products: they include; selective hydrolysis of ß-lg by papain at pH 8.0 (Schmidt and van Markwijk, 1993), by trypsin, α-chymotrypsin or protease from *Aspergillus, Bacillus subtilis* or *Actmomyces* species at pH 7.0-9.0 (Kaneko et al., 1990) or by the enzyme thermolysin at high pressure (2,000 atmospheres) (Hayashi et al., 1987) and selective removal of ß-lg from other whey proteins by affinity chromatography using a column containing ß-lg coupled to Sepharose 4B (Chiancone and Gattoni, 1993).

Minor Whey Proteins. Whey contains a number of proteins that are of biological or pharmaceutical interest, i.e. perhaps 60 indigenous enzymes, vitamin-binding proteins, metal-binding proteins, immunoglobulins, various growth factors and hormones. It is possible that many of these proteins will eventually find commercial application as isolation procedures are improved but at present four are of commercial interest, viz lactoperoxidase, lactotransferrin, immunoglobulins and glycomacropeptide.

Lactoperoxidase. Lactoperoxidase (LPO) is a broad specificity peroxidase present at high concentrations in bovine milk but at low levels or not at all in human milk.

LPO has attracted considerable interest since it has been shown to be involved in the antibacterial activity of various secretions. In milk the antibacterial system consists of LPO, H_2O_2 and ⁻SCN. Commercial interest in LPO derives from activation of the indigenous enzyme for cold sterilization of milk or in the mammary gland to protect against mastitis, and addition of isolated LPO to calf or piglet milk replacers to protect against enteritis, especially when the use of antibiotics in animal feed is not permitted.

LPO is positively charged or cationic at neutral pH and can be isolated from milk or whey by cation exchange chromatography (Paul et al., 1980) which has been scaled up for industrial application (Prieels and Peiffer, 1986). Cation exchangers absorb LPO together with lactotransferrin (Lf) which is also cationic at neutral pH. The LPO and Lf can be eluted together or separately (Yoshida and Ye-Xiuyun, 1991; Kussendrager, 1993). LPO has been isolated from whey by gel filtration using Sephracyl S-200 and hydrophobic chromatography on Butyl Toyopearl 650 M (Yoshida, 1988).

Lactotransferrin. The transferrins are a group of specific metal-binding proteins, the best characterized of which are: serotransferrin, ovotransferrin and lactotransferrin.

Human colostrum and milk contains 6-8 g/l and 2-4 g/l lactotransferrin, respectively, representing ~ 25% of the total protein in the latter; bovine colostrum and milk contain 1.0 and 0.02-0.35 g/l, respectively (Reiter, 1985a, b). Because the concentration of Lf in human milk is considerably higher than that in bovine milk, there is considerable interest in supplementing bovine milk-based infant formulae with bovine Lf. Bovine Lf has also been considered for use in food as an antiseptic or bacteriostatic, in feed as a growth promoter and in medicine as a chemical mediator or iron supplier.

Lfs have been isolated by laboratory scale procedures from the milks of several species and some of the isolation procedures have industrial scale potential. As stated above Lf is cationic at neutral pH and under appropriate conditions cation exchange resins simultaneously bind Lf and LPO and procedures have been patented for their separate recovery. In many cases the basic strategy is the use of whey derived from cheesemaking as the raw material, so that once Lf is isolated the whey can be used for production of other conventional products. In this way, the cost of the raw material for Lf recovery is negligible and only the equipment and processing cost have to be considered. Other methods reported in the literature for the isolation of Lf include gel filtration (Al-Mashiki and Nakai, 1987b) and various affinity chromatographic procedures using immobilized heparin (Bläckberg and Hernell, 1980), ferritin (Pahud and Hilpert, 1976), triazine dyes (Shimazaki and Nishio 1991) and monoclonal antibodies (Kawakami *et al.*, 1987).

Immunoglobulins. Immunoglobulins (Igs), one of the principal defence mechanisms of the body, are present in the mammary secretions, especially colostrum, of all mammalian species. Bovine colostrum protein contains $\sim 10\%$ Igs but this level decreases to $\sim 0.1\%$ within about a week post-partum.

There are large inter-species differences with respect to the concentration and type of Ig, which are largely a reflection of the two mechanisms, *in utero* or *via colostrum*, by which passive immunity by maternal Igs is transferred to the neonate. In general, the milks and colostra of those mammalian species (e.g. human, rabbit) that transfer passive immunity to the foetus *in utero* contain lower amounts and different ratios of the Ig classes than those species (e.g. cow, horse, pig) that transfer passive immunity *via colostrum* after birth and in which ingestion of.Ig from colostrum is essential for the health of the neonate. There is in fact a third group, i.e. mouse, rat, dog, in which Igs are transferred both *in utero* and *via colostrum*. The milks and colostra of those animals that transfer Ig principally *in utero* contain mainly IgA while those that transfer immunity *via colostrum* do so mainly via IgG. The mammary secretions of those species that transfer Ig both *in utero* and *via colostrum* transfer both IgA and IgG.

The principal aspect of Igs of interest here relates to production of Ig concentrates for feeding to neonatal ruminants and pigs. Large molecules such as Igs can be absorbed from the intestine of young ruminants for about 3 days post-partum. Since ruminants are born without antibodies in their blood they are very susceptible to infection and it is highly desirable, probably essential, that they receive colostrum by suckling or pail-feeding within 6 h post-partum. Maternal Igs appear in the blood of the offspring within hours of suckling and can be detected in its blood for about 4 months thereafter. Ruminants start to secrete their own antibodies within ~ 3 weeks of birth and are immunologically independent thereafter.

In situations where it is not possible to feed colostrum to neonatal ruminants and pigs, an alternative source of Igs is necessary and therefore there is interest in the production of Ig concentrates for this purpose. Calf milk replacers and piglet and lamb supplements enriched with Igs are commercially available.

Although human infants are not able to absorb Igs from the intestine, Igs still play an important defensive role in reducing the incidence of intestinal infection. There is general agreement on the superiority of breast-feeding for healthy full-term infants but it is not possible to breast-feed all full-term infants and frequently impossible to breast-feed preterm or very-low-birth-weight infants, the latter may be fed on banked human milk. However, such infants have high protein and energy requirements which may not be met by human milk and consequently special formulae are needed. The use of bovine Igs in infant formulae raises questions because as stated above, there is a marked difference in the predominant type of Igs in human and bovine milks. In human milk and colostrum, the main Ig type is IgA, while in bovine it is IgG1. In humans the role of Igs is the local defence of the intestinal tract against infections, while in the cow there is also a direct transference of Igs from milk to the blood of the calf in the period immediately post-partum.

In spite of the questions raised there is interest in preparation of products containing bovine Igs for use in infant formulae and for use in clinical treatment of diarrhoeas resistant to conventional treatments; these appear in patients affected by HIV and AIDS. The Ig products used should be directed specifically against the pathogens of interest. This often requires previous immunization of the cows from which the Igs are obtained. In some cases, as for rotavirus, the bovine pathogen is similar enough to the human counterpart, and the immunoglobulins directed against the former are able, at least partially, to inactivate the latter. Therefore, it is possible to obtain a certain protective effect on humans with bovine Igs from animals that have not been specifically immunized, and there are commercial products available. However, although cheese whey could be used as raw material to obtain Igs the amount of Igs present in whey is very low. The need to have specificity against pathogens, means that the ideal raw material would be colostrum obtained from hyperimmunized cows during the first 6-8 days of lactation.

The classical method for preparing Ig is by salting out, usually with $(NH_4)_2SO_4$. This method is effective but expensive and current commercial products are usually prepared by ultrafiltration. A "milk immunological concentrate", prepared by ultrafiltration-diafiltration of acid whey from colostrum and early lactation milk from immunized cows, for use in special infant formulae suitable for low-birth-weight pre-term infants has been described (Hilpert, 1984). The milk, whey or concentrate was never exposed to temperatures above 56°C to avoid inactivation of the Igs, and the preparations were sterilized by membrane filtration. The final product contained ~ 75% protein, 50% of which was Ig, mainly IgG1 and not IgA, which is predominant in human milk. Taniguchi et al. (1990) described a process for the preparation of an Ig containing therapeutic agent for rotavirus infection in animals and humans: skimmed colostrum or colostrum whey was ultrafiltered using a 50,000-150,000 Da nominal molecular weight cut-off membrane to concentrate the physiologically active proteins in the retentate. It was claimed that the yield of Igs in the recovered protein and their physiological activity was high. Other methods for the isolation of Igs, sometimes with Lf use ultrafiltration in combination with ion exchange chromatography (Dubois 1986; Bottomley, 1989), immobilized monoclonal antibodies (Gani et al., 1982) and metal chelate (Al-Mashiki et al., 1988) or gel filtration chromatography (Al-Mashiki and Nakai, 1987b).

Glycomacropeptide. There is commercial interest in the recovery of the glycomacropeptide split-off from casein during enzymatic coagulation and present in rennet cheese and casein wheys at a concentration of ~ 1.2-1.5 g/l. The interest in this peptide stems from its unusual amino acid profile, e.g. it contains no Phe, Tyr, Trp, Lys or ½Cys; the absence of aromatic amino acids makes it very suitable for the nutrition of patients suffering from phenylketonurea. The peptide is also claimed to have a therapeutic effect in treatment of viral infection induced diarrhoea (Nielsen and Tromholt, 1994) and to be useful in the treatment of thrombosis (Drouet et al., 1990).

One method for production of a GMP product involves passage of cheese whey through an ion exchange column, GMP is in the non-adsorbed fraction which is desalted and concentrated (Kawaski and Dosako, 1992). A method described by Marshall (1991) involves the selective adsorption of GMP from whey onto an anion exchange column of Spherosil QMA at controlled pH and ionic strength. Subsequent recovery is achieved by desorption with dilute acid or salt solutions. Another method (Kawasaki et al., 1993) involves acidifying of cheese whey to pH 3.5 prior to ultrafiltration through a 50,000 Da nominal molecular weight cut-off membrane through which GMP permeates. When the permeate is neutralized and ultrafiltered again using the same membrane the GMP is retained and the retentate is dried. Yield was claimed to be 63% and purity 81%.

A further method involves UF in combination with heat processing (Nielsen and Tromholt, 1994).

A method described by Berrocal and Neeser (1991) for GMP recovery involves heating a delactosed whey solution to induce protein precipitation. The supernatant containing GMP is recovered and GMP is isolated from this by ethanol precipitation.

Future Developments

In recent years research attention has been focussed on the fractionation of whey proteins with the objective of producing proteins better suited for particular applications than the crude protein mixtures.

Probably the most exciting area for commercial development now is the isolation and/ or production of biologically active proteins and peptides for use in the health food, medical and pharmaceutical areas. It is highly probable that some of the even more minor biologically active proteins present in whey, e.g. various growth factors, will become commercially available in the medium-term future.

Reports on the genetic engineering of milk proteins (Jimenez-Flores and Richardson, 1988; McKnight et al., 1989; Batt et al., 1990; Richardson et al., 1992; Lee et al., 1993)) are now appearing in the literature and in the future genetic engineering may be a very important method by which to tailor functionality and further expand the range of functional milk protein products.

REFERENCES

Al-Mashikhi, S.A. and Nakai, S., 1987a, Reduction of ß-lactoglobulin content of cheese whey by polyphosphate precipitation, J. Food Sci. 52:1237-1242, 1244.

Al-Mashikhi, S.A. and Nakai, S., 1987b, Isolation of bovine immunoglobulins and lactoferrin from whey proteins, J. Dairy Sci. 70:2486-2492.

Al-Mashikhi, S.A., Li-Chan, E. and Nakai, S., 1988, Separation of immunoglobulins and lactoferrin from cheese whey by chelating chromatography, J. Dairy Sci. 70:2486-2492.

Amundson, C.H., Watanawanichakorn, S. and Hill, C.G., 1982, Production of enriched protein fractions of ß-lactoglobulin and α-lactalbumin from cheese whey, J. Food Proc. Preserv. 6:55-71.

Batt, C.A., Robson, L.D., Wong, D.W. and Kinsella, J.E., 1990, Expression of recombinant bovine ß-lactoglobulin in *Escherichia coli*, Agric. Biol. Chem. 54:949-955.

Berrocal, R. and Neeser, J.R., 1991, Process for production of a kappa-casein glycomacropeptide. European Patent Application EP O 453 782 A1.

Bläckberg, L. and Hernell, O., 1980, Isolation of lactoferrin from human whey by a single chromatographic step, FEBS Lett. 109:180-184.

Bottomley, R.C., 1989, Isolation of an immunoglobulin rich fraction from whey. European Patent Application EP 0320 152 A2.

Breslau, B.R. Cross, R.A. and Goulet J., 1975, Production of a crystal clear, bland tasting, protein solution from cheese whey, J. Dairy Sci. 58:782 (Abstract).

Burgess, K.J. and Kelly, J. 1979, Technical note: selected functional properties of a whey protein isolate, J. Food Technol. 14:325-329.

Chiancone E. and Gattoni, M., 1993, Selective removal of ß-lactoglobulin directly from cows milk and preparation of hypoallergenic formulas: a bioaffinity method, Biotech. Appl. Biochem. 18:1-8.

Cuddigan, N.M., 1991, Fractionation of Whey Proteins by Sodium Hexametaphosphate and Functional Properties of the Resulting Fractions, MSc thesis, National University of Ireland, Cork.

de Wit, J.N. and Bronts, H., 1994, Process for the recovery of alpha-lactalbumin and beta-lactoglobulin from a whey protein product. European Patent Application EP 0 604 864.

de Wit, J.N. and de Boer, R., 1975, Ultrafiltration of cheese whey and some functional properties of the resulting whey protein concentrate, Neth. Milk Dairy J. 29:198-211.

de Wit, J.N., Klarenbeek, G. and de Boer, R., 1978, A simple method for the clarification of whey, in: Proceedings of the International Dairy Congress, Paris, Congrilait, Paris, France, pp 919-920.

de Wit, J.N., Hontelez-Backx, E. and Adamse, M., 1988, Evaluation of functional properties of whey protein concentrates and whey protein isolates. 3. Functional properties in aqueous solution. Neth. Milk Dairy J., 42:155-172.

Drouet, L., Sollier, C. Bal dit, Mazoyer, E., Levy-Toledano, S., Jolles, P. and Fiat, A.M. 1990. Use of the caseino-glycopeptide κ, particularly from cow milk, in the manufacture of a composition and particularly of a medicine, for use in the prevention and treatment of thrombosis, French Patent Application FR 2 646 775 A1.

Dubois, E., 1986, Procédé de séparation de certaines protéines du lactosérum ou du lait, French Patent Application FR 2 605 322.

Fox, K.K., Holsinger, V.H., Posati, L.P. and Pallansch, M.J., 1967, Separation of ß-lactoglobulin from other milk serum proteins by trichloroacetic acid, J. Dairy Sci. 50:1363-1367.

Gani, M.M., May, K. and Porter, P., 1982, A process and apparatus for the recovery of immunoglobulins, European Patent 0 059 598 A1.

Hayashi, R., Kawamura, Y. and Kunugi, S., 1987, Introduction of high pressure to food processing: Preferential proteolysis of ß-lactoglobulin in milk whey, J. Food Sci. 52:1107-1108.

Hayes, J.F., Dunkerley, J.A., Muller, L.L. and Griffin, A.T., 1974, Studies on whey processing by ultrafiltration. II. improving permeation rates by preventing fouling, Aust. J. Dairy Technol. 29:132-140.

Hilpert, H., 1984, Preparation of a milk immunoglobulin concentration from cow's milk, in: Human Milk Banking, A.F. Williams and J.B. Baum, eds., Raven Press, New York, pp 17-28.

Janson, H.V. and Lewis, M.J., 1993, Electrochemical coagulation of whey protein, J. Soc. Dairy Technol. 47:87-90.

Jimenez-Flores, R. and Richardson, T., 1988, Genetic engineer of the caseins to modify the behaviour of milk during processing: a review, J. Dairy Sci. 71:2640-2654.

Kaczmarek, J., 1980, Whey protein separation and processing, in: Proceedings of 1980 Whey Production Conference, American Dairy Products Institute, Chicago, USDA Philadelphia, PA, pp 68-80.

Kaneko, T., Kojima, T., Kuwata, T. and Yamamoto, Y., 1990, Selective enzymatic degradation of beta-lactoglobulin contained in cows milk serum protein, European Patent Application, EP 0 355 399 A1.

Kawakami, H., Shinmoto, H., Dosako, S.I. and Sago, Y., 1987, One step isolation of lactoferrin using immobilized monoclonal antibodies, J. Dairy Sci. 70:752-759.

Kawaski, Y. and Dosako, S., 1992, Process for producing kappa-casein glycomacro-peptides, European Patent Application EP 0 488 589 A1.

Kawasaki, Y., Kawakami, H., Tanimoto, M., Dosako, S., Tomizawa, A., Kotake, M. and Nakajima, I., 1993, pH-Dependent molecular weight changes of κ-casein glycomacro-peptide and its preparation by ultrafiltration, Milchwissenschaft 48:191-196.

Kussendrager, K.D., 1993, Process for isolating lactoferrin and lactoperoxidase from milk and milk products, and products obtained by such process, PCT Int. Patent Application WD 93/13676 A1.

Kuwata, T., Pham, A.M., Ma, C.Y. and Nakai, S., 1985, Elimination of ß-lactoglobulin from whey to simulate human milk protein, J. Food Sci. 50:605-609.

Lee, S.P., Cho, Y. and Batt, C.A., 1993, Enhancing the gelation of ß-lactoglobulin, J. Agric. Food Chem. 41:1343-1348.

Lee, D.N. and Merson, R.L., 1976, Chemical treatments of cottage cheese whey to reduce fouling of ultrafiltration membranes, J. Food Sci. 41:778-786.

Mailliart, P. and Ribadeau-Dumas, B., 1988, Preparation of ß-lactoglobulin and ß-lactoglobulin-free proteins from whey retentate by NaCl salting out at low pH, J. Food Sci. 53:743-745, 752.

Marshall, S.C., 1991, Casein macropeptide from whey, A new product opportunity. Food Res. Quart. 51:86-91.

Matthews, M.E., Doughty, R.K. and Short, J.L., 1978, Pretreatment of acid casein whey to improve processing rates in ultrafiltration, N.Z. J. Dairy Sci. Technol. 13:216-220.

Maubois, J.L., Pierre, A., Fauquant, J. and Piat, M., 1987, Industrial fractionation of main whey proteins, IDF Bull. 212, Brussels, Belgium, pp 154-159.

McKnight R.A., Jimenez-Flores, R., Kang Y., Creamer, L.K. and Richardson, T., 1989, Cloning and sequencing of a complementary deoxyribonucleic acid coding for bovine α_{s1}-

casein A from mammary tissue of a homozygous B variant cow, J. Dairy Sci. 72:2464-2473.

Mirabel, B., 1978, Nouveau procédé d'extraction des protéins du lactosérum. Ann. Nutr. l'Aliment, 23:243-253.

Muller, L.L. and Harper, W.J., 1979, Effects on membrane processing of pretreatments of whey, J. Agric. Food Chem. 27:662-664.

Neilsen, P.M. and Tromholt, N., 1994, Method for production of a kappa-casein glycomacropeptide and use of a kappa-casein glycomacropeptide, PCT International Patent Application WO 94 15952 A1.

Pahud, J.J. and Hilpert, H., 1976, Affinity chromatography of lactoferrin on immobilized ferritin, Protides Biol. Fluids 23:571-574.

Palmer, D.E., 1982, Recovery of proteins from food factory waste by ion exchange, in: Food Proteins, P.F. Fox and J.J. Condon, eds., Applied Science Publishers, London, pp 341-352.

Paul, K.G., Ohlsson, P.I. and Hendriksson, A., 1980 The isolation and some liganding properties of lactoperoxidase, FEBS Lett. 110:200-204.

Pearce, R.J., 1987, Fractionation of whey proteins, IDF Bull. 212, Brussels, Belgium, pp 150-153.

Pearce, R.J., 1983, Thermal separation of ß-lactoglobulin and α-lactalbumin in bovine Cheddar cheese whey, Aust. J. Dairy Technol. 38:144-149.

Pierre, A. and Fauquant, J., 1986, Industrial process for production of purified proteins from whey, Le Lait 66:405-419.

Prieels, J.P. and Peiffer, R., 1986, Process for the purification of proteins from a liquid such as milk, UK Patent Application, GB2 171, 102, A1.

Reiter, B., 1985a, The biological significance of the non-immunoglobulin protective proteins in milk: lysozyme, lactoferrin, lactoperoxidase, in: Developments in Dairy Chemistry - 3 - Lactose and Minor Constituents, P.F. Fox, ed., Elsevier Applied Science, London, pp 281-336.

Reiter, B., 1985b, Protective proteins in milk - biological significance and exploitation, IDF Bull 191, Brussels, Beligum, pp 1-35.

Richardson, T., Oh, S., Jimenez-Flores, R., Kumosenski, T.F., Brown, E.M. and Farrell, H.M.Jr., 1992, Molecular modeling and genetic engineering of milk proteins, in: Advanced Dairy Chemistry - 1 - Proteins, P.F. Fox, ed., Elsevier Applied Science, London, pp 545-577.

Schmidt, D.G. and van Markwijk, B.W., 1993, Enzymatic hydrolysis of whey proteins. Influence of heat treatment of α-lactalbumin and ß-globulin on their proteolysis by pepsin and papain, Neth. Milk Dairy J. 47(1):15-22.

Shimazaki, K.-I. and Nishio, N., 1991, Interacting properties of bovine lactoferrin with immobilized Cibacron Blue F3 G-A in column chromatography, J. Dairy Sci. 74:404-408.

Skudder, P.J., 1985, Evaluation of a porous silica-based ion-exchange medium for the production of protein fractions from rennet and acid whey, J. Dairy Res. 52:167-181.

Slack, A.W., Amundson, C.H. and Hill, C.G., 1986, Nitrogen solubilities of ß-lactoglobulin and α-lactalbumin enriched fractions derived from ultrafiltered cheese whey retentate, J. Food Process. Preserv. 10:19-29.

Stack, F.M., Hennessy, M., Mulvihill, D.M. and O'Kennedy, B.T., 1995, Process for the fractionation of whey constituents, PTC International Patent Application WO 953 4216 A1.

Taniguchi, H., Goto, M., Okamoto, T., Sakuchi, I., Ano, T., Kirihara, O. and Ando, K., 1990, Process for preparing a therapeutic agent for rotavirus infection, European Patent Application EPO 391 416 A1.

Yoshida, S., 1988, Isolation of lactoperoxidase of 89,000 Daltons and a globulin of 81,000 Daltons from milk acid whey, J. Dairy Sci. 71:2021-2027.

Yoshida, S. and Ye-Xiuyun, 1991, Isolation of lactoperoxidase and lactoferrins from bovine milk acid whey by carboxymethyl cation exchange chromatography, J. Dairy Sci. 74:1439-1444.

MODIFICATION OF MUSCLE PROTEIN FUNCTIONALITY BY ANTIOXIDANTS

Youling L. Xiong, Subramanian Srinivasan, and Gang Liu

Food Science Section
Department of Animal Sciences
University of Kentucky
Lexington, KY 40546

INTRODUCTION

Protein oxidation is an important chemical process that occurs widely in biological systems. Recent advances in protein research have led to the recognition that muscle proteins, similar to nucleic acids and lipids, can be modified by oxygen free radicals. Such modifications are implicated in the pathogenesis of a number of diseases and certain physiological processes, including aging, ischemia-reperfusion injury, and protein turnover (Stadtman, 1993; Carney and Carney, 1994). Many cellular enzymes, such as alkaline neutral protease involved in protein metabolism, and a number of membrane transport proteins, are susceptible to active oxygen species, and can be readily inactivated due to oxidative damages. Oxidative modification of proteins can occur as a result of attack by free radicals generated via lipid peroxidation, metal ion-catalyzed oxidative reactions, and enzymatic processes (Halliwell and Gutteridge, 1986; Stadtman and Oliver, 1991; Signorini et al., 1995). Free radical-induced physicochemical changes in proteins include protein polymerization (via condensation of protein free radicals), insolubilization, peptide chain scission, and formation of lipid-protein complex (Schaich, 1980; Hanan and Shaklai, 1995).

In the presence of molecular oxygen, iron (Fe^{2+}) and copper (Cu^{2+}), which are naturally abundant in muscle tissue, are strong catalysts of protein oxidation. Metal-catalyzed protein oxidation is believed to be site-specific or "caged", i.e., only amino acid residues at the metal-binding sites are specific targets (Stadtman and Oliver, 1991). Among the most common sites of metal-catalyzed protein oxidation are the alkaline and the sulfur-containing amino acid residues - His, Arg, Lys, Met, and Cys, as well as Pro. The aromatic amino acids, Trp, Tyr, and Phe, are relatively insensitive to metal ions, presumably because they are not commonly present at the metal-binding sites of proteins. In metal-catalyzed protein oxidation, His residues are converted to Asp or Asn residues; Pro residues to Glu and γ-glutamicsemialdehyde residues; and Lys residues to 2-amino-adipicsemialdehyde residues.

Figure 1 illustrates how a Lys residue in proteins can be specifically oxidized by Fe^{2+} in the presence of H_2O_2.

How do proteins located in the muscle fibrils respond to free radical attack, and how would oxidative modification of proteins change the functional behavior of the affected proteins during processing of muscle foods? These are some of the questions concerning protein oxidation which have not been fully addressed in the literature. Limited research has shown that myofibrillar proteins both *in situ* and in their disengaged form (e.g., fibrils) are highly susceptible to oxidizing lipids and metal ions such as Fe^{2+} and Cu^{2+} (Jarenback and Liljemark, 1975; Smith, 1987; Decker et al., 1993; Srinivasan and Hultin, 1995), as well as oxidized myoglobin (radicals) (Bhoite-Solomon et al., 1992; Hanan and Shaklai, 1995). Proteins which are most affected include myosin and actin. Oxidized myofibrillar proteins usually have increased carbonyl and disulfide contents, and strongly oxidized myofibrillar proteins tend to show decreased solubility and gel-forming ability. Incorporation of proper antioxidants in the protein isolation and purification process has been found to inhibit lipid oxidation with a concomitant improvement in protein gelation characteristics (Wan et al., 1993; Xiong et al., 1993; Kelleher et al., 1994).

Our more recent research effort (Srinivasan and Xiong, 1996; Srinivasan et al., 1996; Liu and Xiong, 1996a,b) has been placed on search for the most effective antioxidant treatments and the identification of conditions which would maximize the beneficial effects of antioxidants during the preparation of surimi-like myofibril concentrates particularly from animal by-products. Thus, to obtain improved protein functionality without compromises of flavors of surimi has been our main objective. Possible mechanisms underlying oxidative modification of protein functionality is also an important part of our present studies.

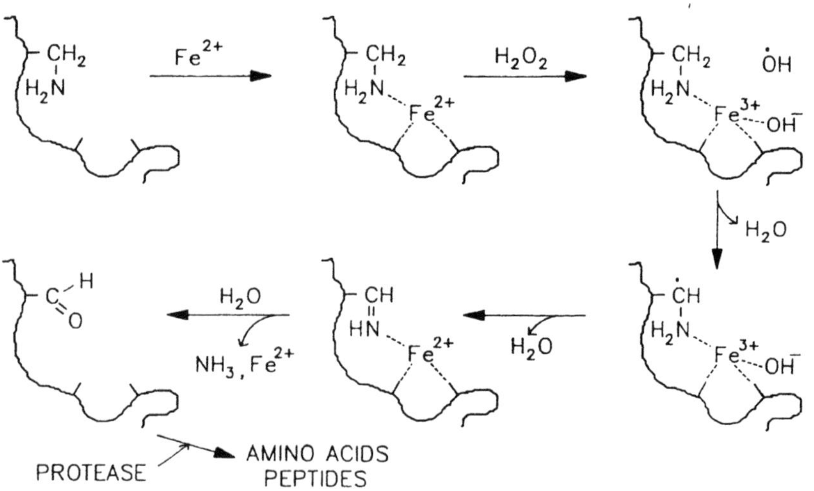

$$NOTE:\ Fe^{2+} + H_2O_2 \rightleftharpoons \left[Fe(OH)_2\right]^{2+} \rightleftharpoons (FeOH)^{3+} + OH^- \rightleftharpoons Fe^{3+} + \dot{O}H + OH^-$$

$$\updownarrow$$

$$(FeO)^{2+} + H_2O$$

Figure 1. Proposed mechanism for the iron-catalyzed site specific oxidation of the ε-amino group of a lysine residue in a protein. (Adapted from Stadtman, 1993. With permission).

MATERIALS AND METHODS

Materials

Fresh beef hearts (24-28 h postmortem) and chicken carcasses (36-48 h postmortem) were obtained from local meat packing and poultry processing plants and used for the preparation of "surimi" (crude myofibrillar proteins) and purified myofibrillar proteins, respectively. Most chemicals, including all antioxidants used, were purchased from Sigma Chemical Co. (St. Louis, MO), Fischer Scientific (Springfield, NJ), and Eastman Kodak (Rochester, NY). All chemicals were at least reagent grade.

Preparation of Beef Heart Surimi

Beef hearts were trimmed of external adipose tissue, caps and valves, and minced with a meat grinder. Surimi was prepared as described by Srinivasan et al. (1996) in a 2-5°C walk-in cooler by washing minced muscle three times with 5-10 vol of the following antioxidant solutions (pH 7.0 in first and second washes, and pH 6.0 in third wash, with a 25 mM sodium phosphate buffer): 1) 0.02% propyl gallate; 2) 0.2% α-tocopherol; 3) 0.2% ascorbic acid; and 4) 0.2% sodium tripolyphosphate. Control washing ("water wash") was done in the absence of any added antioxidant and the first two washes were carried out in non-buffered water. To determine the effect of pH on lipid and protein oxidation, muscle mince was washed three times with 25 mM phosphate buffer at the same pH (5.5, 6.0, or 7.0). Washed surimi was kept on ice up to a week and analyzed periodically for oxidative changes in proteins and alterations in protein functionality.

Preparation of Chicken Myofibrillar Proteins

Myofibrils were isolated from well-trimmed fresh chicken white (breast) and red (leg) muscles in a 2-5°C walk-in cooler as described by Liu and Xiong (1996a). Ground muscle was homogenized and washed four times with 4 vol of a 10 mM potassium phosphate buffer (pH 7.0) containing 0.1 M KCl and the following antioxidants: 1) none (control); 2) 0.02% propyl gallate + 0.2% ascorbic acid; 3) 0.02% propyl gallate + 0.2% sodium tripolyphosphate; or 4) 0.2% ascorbic acid + 0.2% sodium tripolyphosphate. The resultant myofibril pellet, after centrifugation, was washed two additional times in 4 and 8 vol of 0.1 M NaCl solutions. Purified myofibrillar proteins were kept on ice up to a week, and analyzed periodically for oxidative changes in protein chemical and functional properties.

Lipid Oxidation

Thiobarbituric acid-reactive substances (TBARS) as an index of lipid oxidation in beef heart surimi or chicken myofibrillar protein pellets were measured according to MacDonald and Hultin (1987). TBARS was expressed as μg malonaldehyde/g sample.

Protein Oxidation

Protein oxidation was assessed by determining the formation of protein carbonyls (Levine, 1990) and simultaneous loss in soluble free amines (Snyder and Sobocinski, 1975).

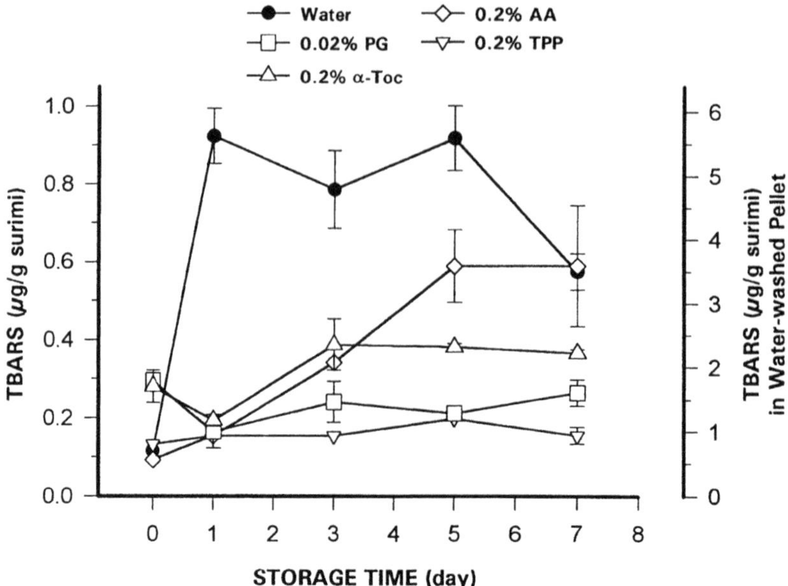

Figure 2. Changes in TBARS in beef heart surimi pellets during storage at 0°C. Surimi was prepared in the presence or absence (water wash) of the following antioxidants: 0.02% propyl gallate (PG), 0.2% α-tocopherol (α-toc), 0.2% ascorbic acid (AA), or 0.2% sodium tripolyphosphate (TPP). (Adapted from Srinivasan et al., 1996. With permission).

Gelation

Beef heart surimi (50 mg/mL protein) and chicken myofibril (20 mg/mL protein) samples were suspended in, respectively, 25 and 50 mM sodium phosphate buffer (pH 6.0) containing 0.6 M NaCl, and heated from 20 to 70°C at 1°C/min to form gels. Small-strain (0.02 amplitude), oscillatory (0.1 Hz) shear measurements of the protein suspensions during the sol → gel transformation were performed using a Bohlin VOR rheometer (Bohlin Instruments, Inc., Cranbury, NJ) as described by Xiong (1993). Gels with a disc geometry (30 mm diameter, 1 mm thickness) were produced between two heated plates. Gel elasticity changes, as measured by shear storage modulus (G'), were constantly monitored during the heating process.

RESULTS AND DISCUSSION

Beef Heart Surimi

Fresh surimi samples prepared from beef heart in the presence of added antioxidants all had an extremely low TBARS value (0.003-0.01 μg/g surimi), indicating that lipid oxidation during surimi preparation was minimal irrespective of the specific antioxidant treatments (Figure 2). Surimi prepared without antioxidants (control or water wash) also had a negligible TBARS content (0.02 μg/g surimi) right after preparation. However, TBARS

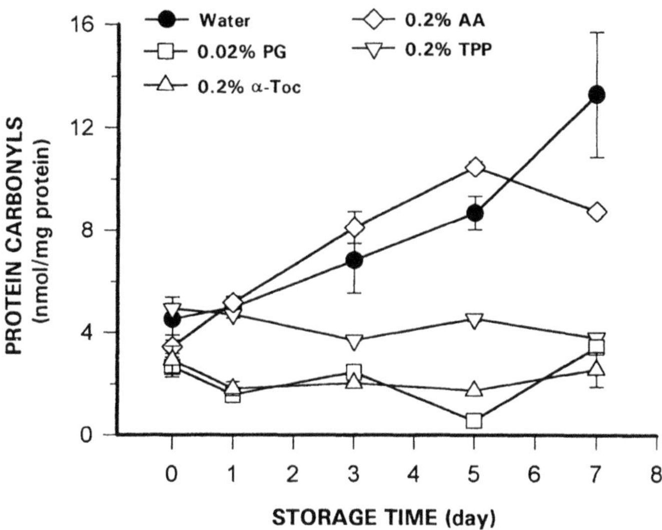

Figure 3. Formation of protein carbonyls in beef heart surimi during storage at 0°C. Surimi was prepared in the presence or absence (water wash) of the following antioxidants: 0.02% propyl gallate (PG), 0.2% α-tocopherol (α-toc), 0.2% ascorbic acid (AA), or 0.2% sodium tripolyphosphate (TPP). (Adapted from Srinivasan et al., 1996. With permission).

in control surimi increased drastically to 0.19 μg/g after 1 day storage. This contrasted sharply with surimi samples washed with either lipid-soluble antioxidants propyl gallate and α-tocopherol, or water-soluble antioxidants tripolyphosphate and ascorbate. Lipid oxidation remained to be essentially completely inhibited during further storage in antioxidant-washed surimi except the surimi washed with ascorbate which exhibited considerable increases after 7 days (Figure 2).

The effectiveness of propyl gallate and α-tocopherol, two potent free radical scavengers, in inhibiting lipid oxidation in minced meat products has been well documented (Greene, 1969; Wan et al., 1993; Kelleher et al., 1994). Tripolyphosphate exerting strong antioxidant activities by chelating metal ions has also been reported (Shahidi et al., 1987). Thus, it is not surprising to see the inhibitory effect of these antioxidants in the surimi pellet even though bovine cardiac muscle is particularly rich in polyunsaturated fatty acids, heme compounds, and mitochondrial oxidative systems. Ostensibly, the strong oxidation of lipids in control surimi resulted from free radical attack which may be catalyzed by metal ions such as Fe^{2+} and Cu^{2+} present in beef heart muscle. The relatively weak antioxidant activity of ascorbate was apparently due to its dual roles. It can act both as an antioxidant and as a prooxidant, depending on its concentration and the environment (Yin et al., 1993). Doba et al. (1985) showed that in the absence of lipid-soluble antioxidants, ascorbate was a good antioxidant for peroxidation initiated in the aqueous phase, but was unable to trap peroxyl radicals in the lipid phase.

Protein carbonyls in water-washed surimi pellet increased almost linearly with storage time (Figure 3). Ascorbate did not inhibit carbonyl formation; on the contrary, it stimulated the formation during storage. However, washing with buffer containing propyl gallate, α-tocopherol, or tripolyphosphate completely suppressed carbonyl production during storage.

Table 1. Effect of pH of washing solution on oxidation of lipids and proteins of fresh and stored (7 days at 0°C) beef heart surimi samples.[1]

pH	TBARS (μg/g surimi)		Protein carbonyl (nmol/mg protein)	
	fresh	7 days	fresh	7 days
5.50	**0.778a**	**2.878a**	**4.54a**	**7.10a**
6.00	**0.552a**	**1.432b**	**4.59a**	**5.09b**
7.00	**0.231b**	**0.648c**	**3.43b**	**3.68c**

[1]Surimi was prepared by washing minced muscle three times with 25 mM phosphate buffer of the same pH (5.50, 6.00, or 7.00). Lipid and protein oxidation was measured as TBARS and carbonyls, respectively. Means with different letters in the same column differ significantly ($P < 0.05$); boldfaced data in the same row within the same parameter group differ significantly ($P < 0.05$). (Data are abstracted from Srinivasan et al., 1996).

Some of the carbonyls detected may be attributed to dehydroascorbate (oxidized L-ascorbate during surimi preparation) which can bind to protein permanently, thereby increasing the amount of measurable protein carbonyls. The results seemed to parallel fairly with those for lipid oxidation (TBARS), suggesting that protein oxidation was probably coupled, to a certain degree, with lipid oxidation in these surimi products. Lipids contained in minced lean beef heart were primarily intramuscular and membranal, and are difficult to remove. The lipid content in the final surimi pellet was as high as 16% on a dry weight basis. Thus, the potential for protein oxidation via oxidized lipids was high in beef heart surimi.

In the absence of added antioxidants, the pH of washing solution had remarkable effects on oxidation of both lipids and proteins. The TBARS contents of fresh surimi samples were relatively small, but samples prepared with pH 5.5 or 6.0 solutions had significantly greater amounts of TBARS than samples prepared at pH 7.0 (Table 1). After 7 days storage, TBARS content increased markedly in surimi prepared at pH 5.5 (from 0.778 to 2.878 μg/g), and moderately in surimi washed at pH 6.0 (from 0.552 to 1.432 μg/g). At pH 7.0, lipid oxidation (TBARS) during storage was largely inhibited. A similar pH effect on the extent of protein oxidation, as indicated by the formation of protein carbonyls, was observed (Table 1). At pH 7.0, carbonyl content of surimi proteins was essentially unchanged during storage. However, protein carbonyls in surimi samples at pH 6.0 and pH 5.5 were more abundant than at pH 7.0 both on day 0 and on day 7 (Table 1). The results further indicate that protein oxidation in beef heart surimi may be associated with lipid oxidation. Compared to pH 5.5, pH 7.0 facilitated the removal of water-soluble sarcoplasmic components. Yang and Froning (1992) presented electrophoretic evidence of substantial removal of heme pigments from mechanically deboned chicken meat at pH 7. Thus, the stronger antioxidative effect of the high-pH washing solution was likely due to the elimination of prooxidative heme compounds from the minced heart tissue. Caughley and Watkins (1985) also reported that high concentrations of H^+ (low pH) favored oxidation of oxymyoglobin to metmyoglobin, resulting in the formation of active oxygen species, which in turn, would catalyze lipid and protein oxidation.

Salted (0.6 M NaCl) surimi samples, irrespective of washing methods, all formed a gel upon heating to above 50°C (Figure 4). The increase in the storage modulus (G') value in the 50-55°C range can be attributed to head-head association of myosin, which led to the formation of weak gel networks. The drop in G' after 55°C is not clearly understood, but may be related to precipitation of protein aggregates or temperature-dependent kinetic variations in protein-protein interactions (Xiong and Blanchard, 1994). The steady increase

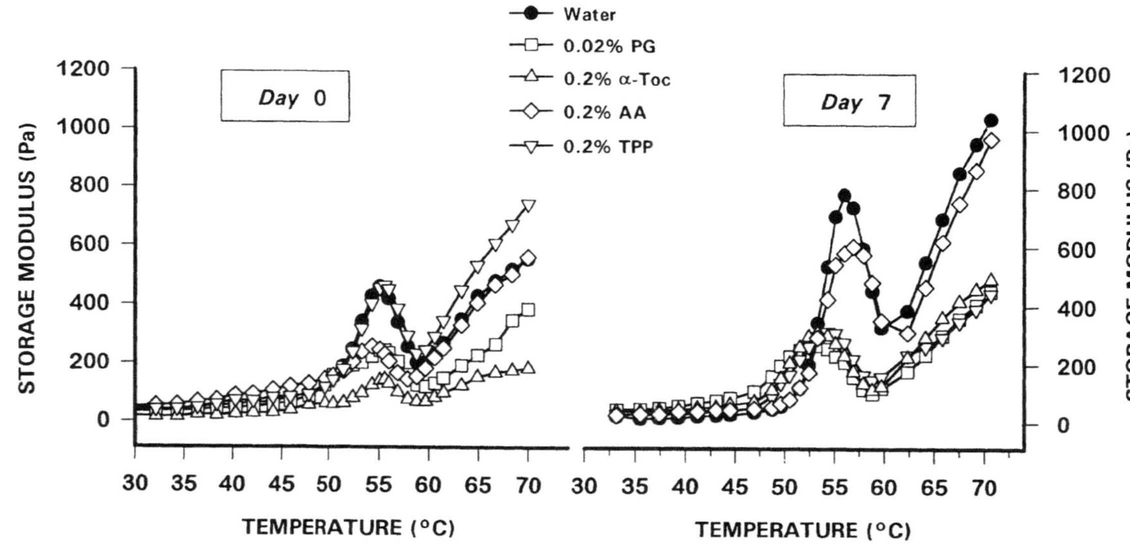

Figure 4. Rheograms of fresh (0 day) and stored (7 days at 0°C) surimi during thermally induced gelation. Surimi was prepared in the presence or absence (water wash) of the following antioxidants: 0.02% propyl gallate (PG), 0.2% α-tocopherol (α-toc), 0.2% ascorbic acid (AA), or 0.2% sodium tripolyphosphate (TPP), and was suspended in 0.6 M NaCl, pH 6.0 before gelation. (Adapted from Srinivasan and Xiong, 1996. With permission).

in G' after 60°C was indicative of formation of a permanent gel network whose elasticity increase with temperature probably reflects both increases in the number of cross-links and "packing" of additional myofibrillar proteins that are denatured at high temperatures. These denatured protein molecules may fill in or deposit onto the existing protein matrices, thereby increasing the mechanical strength of the gel. Regardless of the exact mechanism of gelation, water- and tripolyphosphate-washed fresh surimi (0 day) produced higher peak and final G' values than surimi washed with propyl gallate, ascorbate, or α-tocopherol. Surimi prepared with α-tocopherol formed the weakest gel. Interestingly, after 7 days of storage on ice, gel strength of propyl gallate- and α-tocopherol-washed surimi increased very little, but gel strength of water- and ascorbate-washed surimi increased sharply (Figure 4).

In a previous report, Bhoite-Solomon et al. (1992) demonstrated that intermolecular disulfide bonds of myocardial myosin could form when the myosin was incubated with myoglobin at 37°C. Our present study indicates that interactions between myofibrillar proteins during surimi preparation can occur via non-disulfide bonds as well. Preliminary electrophoretic analysis of washed beef heart surimi showed that both disulfide and non-disulfide covalent bonds between myosin molecules were generated during washing and storage. When G' values of the gels were plotted against protein carbonyls, some relationship between them was observed (Figure 5). Regression analysis revealed a poor correlation between gel G' and protein carbonyl when the extent of oxidation was low in surimi, as was typical of 0-day samples (r=0.37 for peak G'; r=0.44 for final G'). However, correlation between gel G' and protein carbonyl was strong after surimi was stored for 7 days (r=0.99 for both peak and final G'), largely owing to the increase in

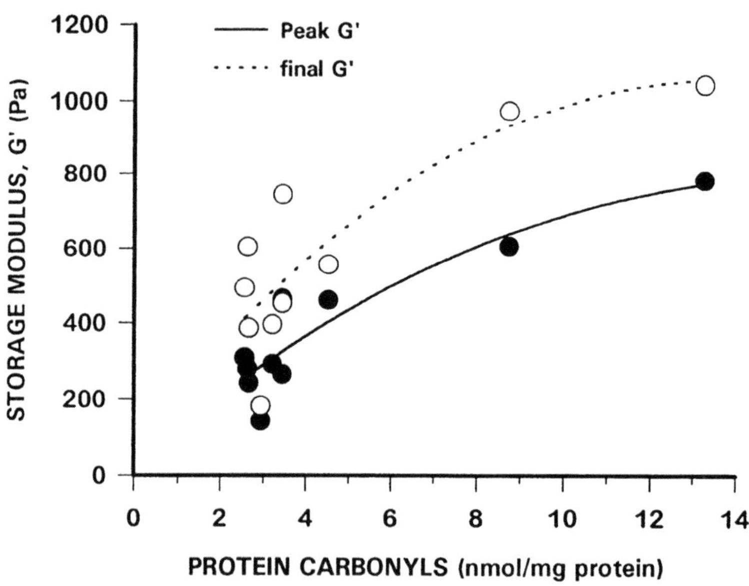

Figure 5. Apparent relationship between protein carbonyl of beef heart surimi and shear storage modulus (G') of its thermally induced gels. Data represent values from all washing treatments, i.e., without (water wash) or with antioxidants (0.02% propyl gallate; 0.2% α-tocopherol; 0.2% ascorbic acid; or 0.2% sodium tripolyphosphate), and from samples stored for 0 or 7 days at 0°C.

protein and(or) protein-bound carbonyls in water- and ascorbate-washed surimi pellets. The overall correlation between protein carbonyl and peak G' (r=0.92) or final G' (r=0.84) of gels was also high. The beneficial effect of ascorbate on kamoboko (a fish surimi-based product) had been well-studied and attributed to polymerization of proteins via dehydroascorbate-catalyzed disulfide bond formation (Lee et al., 1992; Nishimura et al., 1992). Evidence obtained from our study suggests that the mechanism involved in the effect of ascorbate probably had to do with increases in protein-bound carbonyls. It is assumed that protein-bound carbonyls can interact with free amines, producing covalent bonds or cross-links between peptides, thereby increasing the rigidity of the surimi gel network. This postulation is in line with the observation that baking quality of wheat flour dough improves as a result of formation of condensed gluten products from dehydroascorbic acid and free amino groups in the protein (Elkassabany and Hosenet, 1980). There was a lack of correlation between G' of the gels and TBARS even for stored surimi, probably because degradation of TBARS occurred or TBARS reached an equilibrium before storage was completed (Figure 2).

The large G' value of gels made from ascorbate-washed surimi is of particular significance. On one hand, ascorbate-washing greatly inhibited lipid oxidation (TBARS) and hence, protecting surimi against flavor deterioration or rancidity development. On the other hand, ascorbate markedly enhanced the functional properties as manifested by improved gelling ability of the surimi. Our early work showed that use of the combination of ascorbate, tripolyphosphate, and propyl gallate in beef heart surimi preparation also produced satisfactory results, i.e., a complete inhibition of lipid oxidation, yet, a marked improvement in gel-forming and meat-binding properties of washed surimi (Xiong et al., 1993). Despite all the studies conducted so far aimed at modifying protein functionality by using antioxidants, the exact relationship between specific protein functionalities (gelation, emulsification, water-binding, etc.) and the degree of oxidative changes in proteins (carbonyls, disulfides, etc.) remains obscure, and warrants further investigation.

Chicken Myofibrillar Proteins

Poultry meat is increasingly being utilized in the preparation of further processed, ready-to-eat products. Our previous studies have shown that proteins from turkey skeletal muscle are very susceptible to oxidative deterioration and functionality losses when exposed to oxidized lipids or iron and copper in the presence of ascorbate (Decker et al., 1993). To determine how lipid and protein oxidation might be prevented and how this could influence muscle protein functionality, we prepared myofibrillar proteins from chicken skeletal muscle by using combined antioxidants in the washing media. Chicken breast (a white muscle) and leg (a red muscle) were used to build a model system because the two muscle types differ in the concentration of anti- and prooxidants, and thus, would speculatively respond to antioxidants differently during protein purification.

Incorporation of combined antioxidants propyl gallate and ascorbate, or propyl gallate and tripolyphosphate, markedly inhibited lipid oxidation (TBARS) in fresh and stored (2 and 6 days) protein samples for both breast and leg muscle (Table 2). However, the addition of ascorbate together with tripolyphosphate in the washing buffer promoted lipid oxidation. Thus, scavenging of free radicals was crucial for stopping lipid oxidation, since only in the presence of propyl gallate was the oxidation inhibited. The effect of ascorbate and tripolyphosphate differed on chicken myofibril concentrate (prooxidation) than on beef heart surimi (antioxidation), indicating that modes of action of these water-soluble antioxidants were different in the two muscle systems. Compared to chicken skeletal muscle, beef heart muscle is particularly high in mitochondrial lipids, heme pigments (e.g., cytochrome C), and

Table 2. Oxidation of lipids and proteins and changes in gelation properties of antioxidant-washed chicken myofibril pellets during storage at 0°C.[1]

Parameter	Breast			Leg		
	0	2	6 days	0	2	6 days
TBARS (μg/g pellet)						
Control	**0.154c**	**0.182b**	**0.283a**	**0.100b**	**0.164b**	**0.580a**
PG + AA	0.036	0.084	0.082	0.006	0.045	0.076
PG + TPP	0.058	0.045	0.065	0.042	0.050	0.050
AA + TPP	**0.130b**	**0.156b**	**0.438a**	**0.089b**	**0.551a**	**1.291a**
Mean	0.095	0.117	0.217	0.059	0.203	0.499
Protein carbonyl (nmol/mg protein)						
Control	1.570	1.450	2.053	1.208	0.966	2.536
PG + AA	1.087	2.053	2.416	0.242	1.328	1.690
PG + TPP	1.087	1.328	1.328	1.449	1.570	1.570
AA + TPP	0.362	1.328	1.570	0.483	1.450	2.294
Mean	1.026	1.540	1.840	0.846	1.330	2.023
Free amine (nmol/mg protein)						
Control	10.720	9.886	7.197	11.440	8.613	7.386
PG + AA	9.508	7.424	7.046	10.378	12.576	10.036
PG + TPP	7.348	7.765	3.182	7.954	9.280	5.796
AA + TPP	9.621	10.379	6.553	10.492	11.174	8.485
Mean	9.300	8.860	5.990	10.070	10.410	7.930
Peak G' (Pa)						
Control	367a	301b	275b	154a	108b	83c
PG + AA	305a	244b	170c	131a	100b	73c
PG + TPP	369a	298b	217c	161a	144a	128b
AA + TPP	423a	328b	293c	168a	133b	106c
Mean	366	293	239	153	121	97
Final G' (Pa)						
Control	472c	503b	535a	224b	240b	263a
PG + AA	367b	424a	451a	193	202	244
PG + TPP	413	419	354	201	209	221
AA + TPP	502	503	533	187b	187b	204a
Mean	438	462	468	201	209	233

[1]Means within the same row in the same muscle type differ significantly (P < 0.05) only when they bear different letters. Boldfaced data in the same column within the same parameter group differ (P < 0.05) from non-boldfaced data. No significant (P > 0.05) difference in chemical traits (TBARS, protein carbonyl, free amine) was found between breast and leg muscle myofibril pellets for the same storage periods; however, both peak and final storage moduli (G') of breast gels differ significantly (P < 0.05) from those of leg gels for all storage periods. PG = propyl gallate; AA = ascorbic acid; TPP = sodium tripolyphosphate. (Data are abstracted from Liu and Xiong, 1996a,b).

oxidative enzymes. Hence, oxidative pathways for cardiac lipids may deviate from those of skeletal muscle lipids. As forementioned, ascorbate is a dual-function agent, capable of both inhibiting and promoting oxidation depending on its concentration and the environment. In fact, on a similar protein concentration basis, the content of TBARS in beef heart surimi was 2 to 3-fold higher than in chicken skeletal muscle protein pellets. There was no significant difference in the TBARS content between breast and leg myofibrillar protein samples for each of the antioxidant treatments, despite the fact that chicken leg muscle used in our experiments contained considerably larger amounts of lipids (2.3 g/100 g) and iron (1.0 mg/100 g) than chicken breast muscle (lipids: 0.6 g/100 g; iron: 0.7 mg/100 g). The mechanisms are not clear, but could be related to the different concentrations of pro- and antioxidants naturally present in white and red muscle tissues.

Protein samples washed without antioxidant (control) or with ascorbate experienced large increases in carbonyls during storage when compared to samples washed with propyl gallate and tripolyphosphate. The results were consistent with the observation on beef heart surimi, and suggest that carbonyl is a sensitive index of protein oxidation in muscle foods, and it could be related to lipid oxidation. Changes in free amines in the myofibril pellets during storage were variable, and appeared to be independent of antioxidant treatments (Table 2). Nevertheless, the disappearance of free amines, on average, coincided with the formation of protein carbonyls, suggesting that some of the carbonyls might be produced via free radical-catalyzed deamination of lysyl or arginyl residues as proposed by Stadtman and Oliver (1991).

The rheogram of thermally induced chicken myofibrillar protein gels resembled that of beef heart surimi gels, and a transition peak (45-55°C) in the storage modulus curve was identified. In beef heart surimi, gel strength (both peak and final G') was strongly correlated with protein carbonyls. This was not entirely the case for chicken breast and leg myofibrillar proteins. In fact, peak G' values of chicken protein gels generally decreased during storage, while the protein carbonyl content increased (Table 2). However, the final G' values of chicken protein gels did increase during storage of the protein pellets, corresponding to increases in protein carbonyls. Presumably, oxidation occurring during storage of the myofibril pellets led to physicochemical modifications of proteins in such a way that protein-protein interactions in the 45-55°C temperature range, the initial stage of protein network formation, was suppressed. This suppression, however, seemed to be conducive to the development of stronger gel network after the temperature increased to 60°C or higher. Similarly, there was a negative relationship between the peak G' of the gels and TBARS of the protein pellets, but a positive relationship between the final G' of the gels and the TBARS. Among all washing treatments on breast muscle proteins, gels made from control-washed proteins and proteins washed with ascorbate and tripolyphosphate showed the largest G' values, although the amount of protein carbonyls in control proteins and ascorbate and tripolyphosphate-washed proteins was not always high. The formation of protein carbonyls is only one of the oxidative changes involved during protein oxidation. There are certain cases in which oxidative changes in proteins can take place without forming carbonyl or animo derivatives, e.g., oxidation of histidine and proline and the loss of sulfhydryl groups (to form S-S bonds) (Takenaka et al., 1991). These possible additional oxidative changes could also lead to alterations in protein functionalities. Recently, Hanan and Shaklai (1995) reported that myoglobin radicals could be generated in the presence of H_2O_2. These free radicals were capable of catalyzing cross-linking between myosin molecules due to electron transfer via tyrosine residues. Implication of such oxidative processes in protein functionality is evident, and could occur during the isolation of chicken myofibrillar proteins.

With the same purification procedure and conditions, breast myofibrillar proteins formed stronger gels than leg myofibrillar proteins. The equal susceptibility of chicken breast and leg proteins to oxidation and their similar responses to the antioxidant treatments suggest that functional discrepancies between white (fast-twitch) and red (slow-twitch) muscle proteins found in the present as well as in numerous previous studies may be irrelevant to oxidative modification of the proteins which can occur *in situ* or during protein isolation and purification. Instead, they can be ascribed to physicochemical differences which are specific to muscle fiber types and dependent on isoforms of myosin (Xiong, 1994).

CONCLUSION

Lipid and protein oxidation occur during preparation of protein concentrates from both cardiac and skeletal muscles of meat animals. Oxidative changes, such as formation of protein carbonyls, can lead to significant alterations in protein functionalities including gelation. By incorporation of antioxidants in the washing process, it is possible to control oxidative changes in lipids and proteins to obtain a broad range and varying magnitude of protein functionalities. Oxidative processes involved in muscle washing are quite complex, and apparently not limited to deamination or formation of carbonyls. For instance, protein gel-forming ability can be modified as a result of cross-linking between protein free radicals due to hydrogen abstraction and oxidation of sulfhydryl groups. Extreme protein oxidation render proteins insoluble by forming large protein aggregates and particulates, and this usually result in losses in functionalities. Therefore, antioxidative or controlled oxidative processes should be designed to allow limited (mild) protein oxidation, and at the same time, minimizing lipid oxidation and generation of oxidative off-flavors. The exact relationship between the extent of protein oxidation and protein functional properties in muscle food processing requires more detailed and extensive examinations. These should include the identification and delineation of various physicochemical changes induced by pro- and antioxidant treatments which are most critical to protein functionalities.

REFERENCES

Bhoite-Solomon, V., Kessler-Icekson, G., and Shaklai, N., 1992, Peroxidative crosslinking of myosins, *Biochem. Int.* 26:181-189.

Carney, J.M., and Carney, A.M., 1994, Role of protein oxidation in aging and in age-associated neurodegenerative diseases, *Life Sci.* 55:2097-2103.

Caughley, W.S., and Watkins, J.A., 1985, Oxy radical and peroxide formation by hemoglobin and myoglobin, in: *CRC Handbook of Methods for Oxygen Radical Research*, R.A. Greenwald, ed., CRC Press, Boca Raton, FL.

Decker, E.A., Xiong, Y.L., Calvert, J.T., Crum, A.D., and Blanchard, S.P., 1993, Chemical, physical and functional properties of oxidized turkey white muscle myofibrillar proteins, *J. Agric. Food Chem.* 41:186-189.

Doba, T., Burton, G.W., and Ingold, K.U, 1985, Antioxidant and co-antioxidant activity of vitamin C. The effect of vitamin C, either alone or in the presence of vitamin E or a water-soluble vitamin E analogue, upon the peroxidation of aqueous multilamellar phospholipid liposomes, *Biochim. Biophys. Acta* 835:298-303.

Elkassabany, Y., and Hoseney, R.C., 1980, Ascorbic acid as an oxidant in wheat flour dough. II. Rheological effects, *Cereal Chem.* 57:88-91.

Greene, B.E., 1969, Lipid oxidation and pigment changes in raw beef, *J. Food Sci.* 4:110-113.

Halliwell, B., and Gutteridge, J.M.C., 1986, Oxygen free radicals and iron in relation to biology and medicine: Some problems and concepts, *Arch. Biochem. Biophys.* 246:501-514.

Hanan, T., and Shaklai, N., 1995, The role of H_2O_2-generated myoglobin radical in crosslinking of myosin, *Free Radic. Res.* 22:215-217.

Jarenback, L., and Liljemark, A., 1975, Ultrastructural changes during frozen storage of cod. III. Effects of linoleic acid and linoleic acid hydroperoxides on myofibrillar proteins, *J. Food Technol.* 10:437-452.

Kelleher, S.D., Hultin, H.O., and Wilhelm, K.A., 1994, Stability of mackerel surimi prepared under lipid-stabilizing processing conditions, *J. Food Sci.* 59:269-271.

Lee, H.G., Lee, C.M., Chung, K.H., and Lavery, S.A., 1992, Sodium ascorbate affects surimi gel-forming properties, *J. Food Sci.* 57:1343-1347.

Levine, R.L., Garland, D., Oliver, C.N., Amici, A., Climent, I., Lenz, A.-G., Ahn, B.-W., Shalitel, S., and Stadtman, E.R., 1990, Determination of carbonyl content in oxidatively modified proteins, *Meth. Enzymol.* 186:464-478.

Liu, G., and Xiong, Y.L., 1996a, Storage stability of antioxidant-washed myofibrils from chicken white and red muscle, *J. Food Sci.* 61:(in press).

Liu, G., and Xiong, Y.L., 1996b, Contribution of lipid and protein oxidation to rheological differences between chicken white and red myofibrillar proteins, *J. Agric. Food Chem.* 44:(in press).

McDonald, R.E., and Hultin, H.O., 1987, Some characteristics of the enzymic lipid peroxidation system in the microsomal fraction of flounder skeletal muscle, *J. Food Sci.* 52:15-21, 27.

Nishimura, K., Ohishi, N., Tanaka, Y., and Sasakura, C., 1992, Participation of radicals in polymerization by ascorbic acid of crude actomyosin from frozen surimi of Alaska pollack during a 40°C incubation, *Biosci. Biotech. Biochem.* 56:24-28.

Schaich, K.M., 1980, Free radical initiation in proteins and amino acids by ionizing and ultraviolet radiations and lipid oxidation - part III: free radical transfer from oxidizing lipids, *CRC Crit. Rev. Food Sci. Nutr.* 13:189-244.

Shahidi, F., Rubin, L.J., and Wood, D.F., 1987, Control of lipid oxidation in cooked meats by combinations of antioxidants and chelators, *Food Chem.* 23:151-157.

Signorini, C., Ferrali, M., Ciccoli, L., Sugherini, L., Magnani, A., and Comporti, M., 1995, Iron release, membrane protein oxidation and erythrocyte ageing, FEBS Lett. 362:165-170.

Smith, D.M., 1987, Functional and biochemical changes in deboned turkey due to frozen storage and lipid oxidation, *J. Food Sci.* 52:22-27.

Snyder, S.L., and Sobocinski, P.Z., 1975, An improved 2,4,6-trinitrobenzenesulfonic acid method for the determination of amines, *Anal. Biochem.* 64:284-288.

Srinivasan, S., and Hultin, Herbert, O., 1995, Hydroxyl radical modification of fish muscle proteins, J. Food Biochem. 18:405-425.

Srinivasan, S., and Xiong, Y.L., 1996, Gelation of beef heart surimi as affected by antioxidants, *J. Food Sci.* 61:(in press).

Srinivasan, S., Xiong, Y.L., and Decker, E.A., 1996, Inhibition of protein and lipid oxidation in beef heart surimi-like material by antioxidants and combinations of pH, NaCl, and buffer type in the washing media, *J. Agric. Food Chem.* 44:(in press).

Stadtman, E.R., 1993, Oxidation of free amino acids and amino acid residues in proteins by radiolysis and by metal-catalyzed reactions, *Annu. Rev. Biochem.* 62:797-821.

Stadtman, E.R., and Oliver, C.N., 1991, Metal-catalyzed oxidation of proteins, *J. Biol. Chem.* 266:2005-2008.

Takenaka, Y., Yasuda, H., and Mino, M., 1991, The effect of α-tocopherol as an antioxidant on the oxidation of membrane protein thiols induced by free radicals generated in different sites, *Arch. Biochem. Biophys.* 285:344-350.

Wan, L., Xiong, Y.L., and Decker, E.A., 1993, Inhibition of oxidation during washing improves the functionality of bovine cardiac myofibrillar protein, *J. Agric. Food Chem.* 41:2267-2271.

Xiong, Y.L., 1993, A comparison of the rheological characteristics of different fractions of chicken myofibrillar proteins, *J. Food Biochem.* 16:217-227.

Xiong, Y.L., 1994, Myofibrillar protein from different muscle fiber types: implications of biochemical and functional properties in meat processing, *CRC Crit. Rev. Food Sci. Nutr.* 34:293-320.

Xiong, Y.L., and Blanchard, S.P., 1994, Myofibrillar protein gelation: viscoelastic changes related to heating procedures, *J. Food Sci.* 59:734-738.

Xiong, Y.L., Decker, E.A., Robe, G.H., and Moody, W.G., 1993, Gelation of crude myofibrillar protein isolated from beef heart under antioxidative conditions, *J. Food Sci.* 58:1241-1244.

Yang, T.S., and Froning, G.W., 1992, Effects of pH and mixing time on protein solubility during the washing of mechanically deboned chicken meat, *J. Muscle Foods* 3:15-23.

Yin, M.C., Faustman, C., Riesen, J.W., and Williams, S.N., 1993, Alpha-tocopherol and ascorbate delay oxymyoglobin and phospholipid oxidation *in vitro*, *J. Food Sci.* 58:1273-1276.

THE SEED STORAGE PROTEINS OF QUINOA

Chris Brinegar

Department of Biological Sciences
San Jose State University
One Washington Square
San Jose, CA 95192

INTRODUCTION

Quinoa (*Chenopodium quinoa*) is a member of the Chenopodiaceae, or goosefoot, family of dicotyledenous plants (Figure 1). Indigenous to the Andean region of South America, cultivated quinoa is grown primarily in the cool, semi-arid altiplano of Bolivia and Peru but is also adaptable to valleys and coastal foothills. Quinoa has been a staple food item in Andean culture since 3000 B.C and is sometimes referred to as Inca rice. Quinoa seeds can be boiled like rice or ground into a flour. The leaves are also edible, having a mild spinach-like flavor. It is still a major dietary component among the present-day Quechua and Aymara Indians (Wood, 1985).

Quinoa has been promoted as a highly nutritious and agriculturally adaptable crop having potential for worldwide cultivation (Risi and Galwey, 1984; National Research Council, 1989). Quinoa's resistance to salty and alkaline soil conditions, drought, and frost allows it to be grown in marginally arable land in cool climates having as little as four inches of annual precipitation (Wood, 1985) although 10-15 inches is recommended (Johnson and Croissant, 1989). Depending on growing conditions, quinoa plants can attain heights of 3-6 feet with thousands of small seeds arranged in compact clusters on a panicle. Seed yields under normal conditions range from 1000-3000 kg/ha (Rea et al., 1979). Quinoa "seeds" are actually fruits composed of a thin, saponin-rich pericarp (seed coat) surrounding a starchy, disc-shaped perisperm which is encircled by an oil and protein-rich embryo (Varriano-Marston and DeFrancisco, 1984). Seeds must be washed to remove the saponins (oleanolic acid and hederagenin) which have a bitter taste and slight hemolytic effects (Burnouf-Radosevich and Delfel, 1984; Burnouf-Radosevich, 1988).

Grown commercially in the U.S. since the mid-1980's, quinoa has been distributed successfully in the health food market as bulk grain or incorporated as a flour into cereals, breads, and pastas. Quinoa seeds have a relatively high protein content which averages approximately 12.6% of seed dry weight (Cardoza and Tapia, 1979), and their overall amino acid

Figure 1. Quinoa seed head (A), leaf (B), and tap root (C) of a lowland ecotype from Chile grown near sea level in Santa Cruz County, California. Ecotypes of quinoa seeds (D) show variations in color. The commercial U.S. grown quinoa seed ("Ancient Harvest") used for protein isolation is at far left.

composition rivals the FAO standard in essential amino acid profile (Wood, 1985). Protein quality studies have shown quinoa to be an excellent source of dietary protein, being nutritionally equivalent to casein (Mahoney et al., 1975).

Until recently, little was known about the storage proteins in quinoa seeds. Polymorphisms in quinoa seed protein electrophoretic patterns were investigated as genetic markers for quinoa breeding (Burnouf-Radosevich, 1988; Fairbanks et al., 1990), but no studies were undertaken to characterize individual quinoa storage proteins until Brinegar and Goundan (1993) reported the isolation, subunit structure, N-terminal sequence, and amino acid composition of chenopodin, the 11S-type storage protein of quinoa. Recently, the other major storage protein of quinoa seeds, a high-cysteine 2S fraction, was isolated (Brinegar et al., 1996).

In this presentation the isolation, structure, properties, and amino acid compositions of these two classes of quinoa seed storage proteins will be reviewed, and suggestions will be made regarding the potential dietary and functional uses of quinoa proteins in both developed and developing countries.

MATERIALS AND METHODS

Extraction of protein from defatted quinoa seed flour, chromatography, sodium docecyl sulfate polyacrylamide gel electrophoresis (SDS-PAGE), N-terminal sequencing of the basic subunit, and amino acid analysis of the 11S protein (chenopodin) have been described previously (Goundan, 1992; Brinegar and Goundan, 1993). The isolation and amino acid analysis of the quinoa 2S seed protein is described by Brinegar et al. (1996). A detailed flow chart of the 11S and 2S protein isolation protocols is shown in Figure 2.

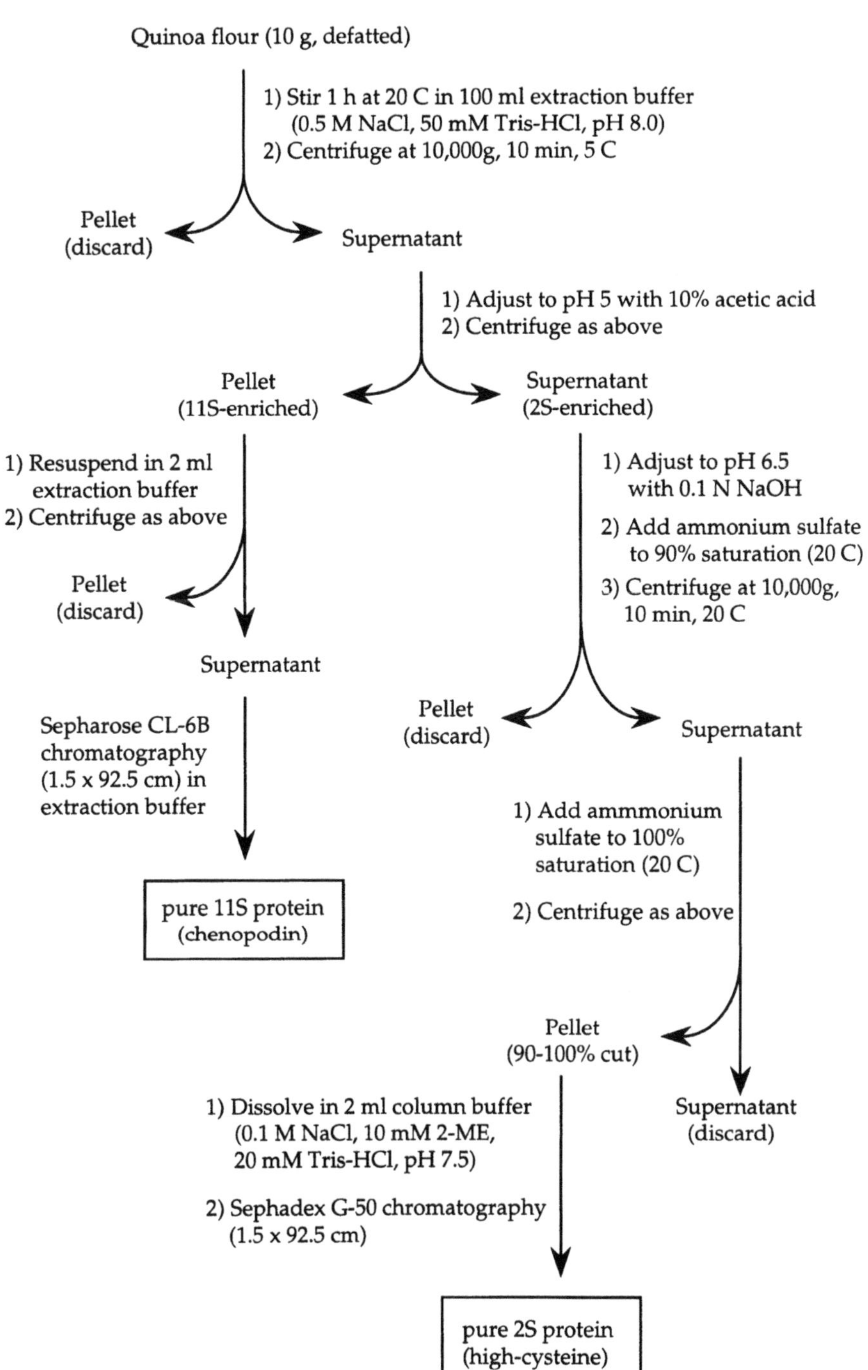

Figure 2. Isolation flow chart for quinoa 11S and 2S seed storage proteins. (Adapted with permissionfrom Brinegar and Goundan, 1993 and Brincgar et al., 1996. Copyright American Chemical Society.)

RESULTS

Seed Protein Extraction

High concentrations of sodium chloride are required for optimal extraction of the major quinoa seed storage proteins (Figure 3). A linear relationship was observed between extractable protein and salt concentration between 0.1 and 0.5 M NaCl with maximal extraction achieved at 1.0 M NaCl. Even without salt, the pH 8.0 buffer alone could extract approximately two-thirds of the amount of protein extracted at 1.0 M NaCl. Without salt there was less efficient extraction of a group of 8-9 kDa polypeptides (the 2S proteins), but, in general, the polypeptide patterns seen by SDS-PAGE were very similar regardless of the salt concentration used (Goundan, 1992). The 11S protein family (chenopodin) is represented by the polypeptides between 22 and 39 kDa.

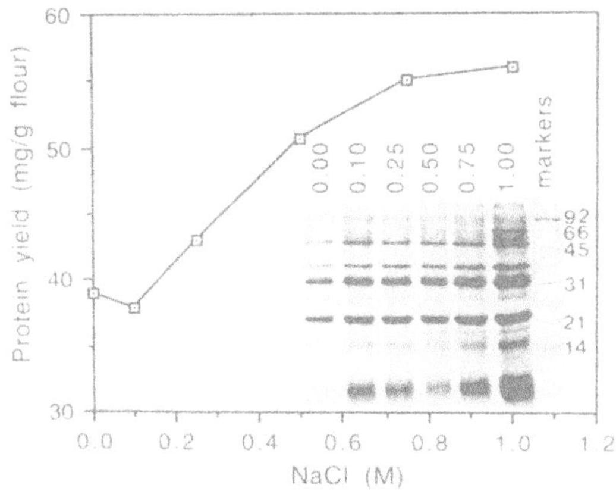

Figure 3. Effect of NaCl concentration on quinoa seed protein extraction. Protein was extracted from 1 g lots of defatted quinoa flour for 1 hr at room temperature in 50 mM Tris-HCl with varying concentrations of NaCl. After a 10,000*g* centrifugation, the supernatant protein was determined by the Bradford method . SDS-PAGE inset: Patterns of quinoa polypeptides extracted at various NaCl concentrations. Molarity of NaCl is indicated above the lanes; molecular weights of the markers (in kDa) are at right.

Isolation of Chenopodin

As shown in the flow chart (Figure 2), chenopodin can be isolated easily from a salt extract by a simple two-step procedure. Acidification of the extract to pH 5.0 quantitatively precipitates chenopodin. Following the dissolution of the precipitate at pH 8.0, Sepaharose CL-6B chromatography (Figure 4) removes high and low molecular weight UV-absorbing (non-proteinaceous) material along with the major contaminating protein, a 50 kDa polypeptide (Brinegar and Goundan, 1993).

In the unreduced form, denatured chenopodin exists as 55-62 kDa heterodimers of A and B subunits which can be separated by reduction of their disulfide bonds (Figure 4 SDS-PAGE inset, lanes 2 and 3) (Goundan, 1992). The estimated size of the native protein is approximately 320 kDa (Brinegar and Goundan, 1993), suggesting that chenopodin is composed of six heterodimers. This quaternary structure is common to all 11S seed storage proteins (Derbyshire et al.,1976).

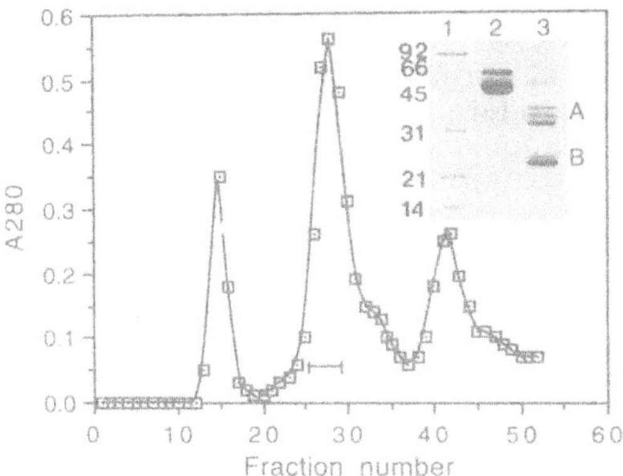

Figure 4. Purification of the quinoa 11S protein. Graph: Sepaharose CL-6B chromatography of the redissolved protein precipitated from a quinoa flour extract at pH 5.0. Bar indicates fractions containing the pure 11S protein. (Adapted with permission from Brinegar and Goundan, 1993. Copyright American Chemical Society.) SDS-PAGE inset: Purified 11S protein (chenopodin). Lane 2, unreduced. Lane 3, reduced. The A and B subunit groups are indicated at right.

Isolation of the 2S High-Cysteine Protein

In contrast to the insolubility of chenopodin at pH 5.0, the 2S class of quinoa proteins are extremely soluble. Therefore, the 2S protein remains in the supernatant after pH 5.0 treatment of the extract (Figure 5 SDS-PAGE inset, lanes 2 and 3). A subsequent 90-100% ammonium sulfate cut precipitates the 2S protein along with a 15 kDa contaminant (lane 4) which can be removed by Sephadex G-50 chromatography (Figure 5 chromatogram). Selected fractions from the main protein peak contain the purified 2S protein (lane 5) (Brinegar et al., 1996).

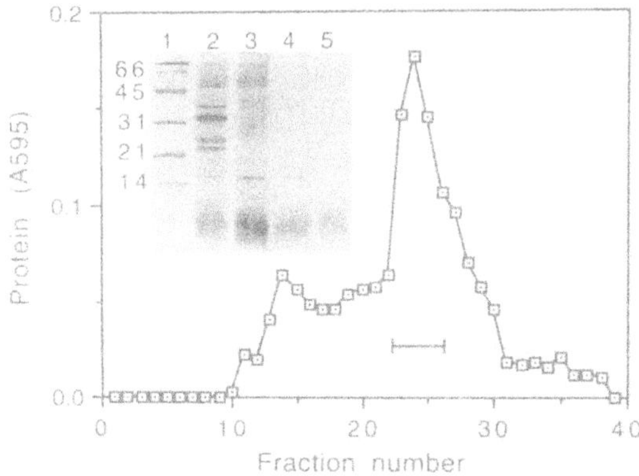

Figure 5. Purification of the quinoa 2S protein. SDS-PAGE inset: Lane 1, markers; lane 2, flour extract; lane 3, supernatant of flour extract after the pH 5.0 precipitation of chenopodin; lane 4, 90-100% ammonium sulfate cut of the pH 5.0 supernatant; lane 5, purified 2S protein after Sephadex G-50 chromatography of the 90-100% ammonium sulfate cut. (Adapted with permission from Brinegar et al., 1996. Copyright American Chemical Society.) Bar on the chromatogram indicates the Sephadex G-50 fractions containing the pure 2S protein.

Amino Acid Compositions of Quinoa Seed Proteins

Amino analyses of purified chenopodin and the 2S protein are shown in Table 1 with the residues arranged in groups of the acid/amide, neutral, hydroxyl, sulfur, hydrophobic, and basic amino acids, respectively.

The amino acid composition of chenopodin is typical of other 11S seed storage proteins (Derbyshire et al.,1976). It is high in arginine (9.7 mol %) and the acid/amide amino acids (29.3 mol %), low in the sulfur amino acids (2.5 mol %), with medium amounts of the other amino acids.

The 2S protein composition is significantly different than that of chenopodin and is striking in the fact that over 60% of its amino acids are either glutamic acid/glutamine (30.8 mol%), cysteine (15.6 mol %), or arginine (15.2 mol %). Based on an average molecular weight of 8.5 kDa, each 2S polypeptide should contain 12 cysteines per molecule.

Table 1. Amino acid compositions[1] of the major quinoa seed storage proteins.

Amino acid	11S protein[2] (chenopodin)	2S protein[3] (high-cysteine)
Glutamic acid/Glutamine	14.8	30.8
Aspartic acid/Asparagine	10.5	5.3
Glycine	8.7	7.4
Alanine	5.3	1.4
Proline	4.9	4.5
Threonine[4]	3.9	0.4
Serine[4]	8.9	2.0
Methionine	1.7	0.6
Cysteine[5]	0.8	15.6
Valine[6]	5.4	0.7
Isoleucine	4.9	1.3
Leucine	7.4	2.2
Phenylalanine	4.0	1.2
Tyrosine	2.9	2.9
Lysine	3.2	0.9
Histidine	3.0	7.6
Arginine	9.7	15.2

[1]Values are expressed as mole percent. Tryptophan is not shown due to destruction by acid hydrolysis.

[2,3]Adapted with permission from Brinegar and Goundan, 1993 and Brinegar et al., 1996, respectively. Copyright American Chemical Society.

[4]Corrected for partial degradation during acid hydrolysis.

[5]Determined as carboxymethylcysteine (11S) or pyridylethylcysteine (2S).

[6]The previously reported 11S value of 6.8 (Brinegar and Goundan, 1993) was incorrect.

DISCUSSION

The two major classes of proteins in quinoa seeds, the 11S (chenopodin) and 2S (high-cysteine) proteins, can be very easily separated from each other by taking advantage of the insolubility of the chenopodins at their isoelectric point near pH 5.0. Purification of the two requires only ammonium sulfate precipitation and/or gel filtration chromtography. Both types of proteins are electrophoretically heterogeneous and most likely are encoded by complex families of genes.

The distinctive structural and solubility characteristics of the 11S and 2S proteins suggest that their functional properties should differ markedly. The insolubility of chenopodin under acidic conditions is characteristic of other 11S proteins, some of which (e.g., glycinin from soybeans) are capable of forming strong gels. The quinoa 2S proteins have two properties which could make them useful as a functional proteins: high solubility and numerous cysteine residues. Such high-cysteine proteins might find use as components in hair care products.

Finally, the combination of the two proteins yields an overall amino acid composition rich in most of the essential amino acids. The 2S protein is especially rich in cysteine, histidine, and arginine, all of which humans can synthesize. However, cysteine has a sparing effect on dietary methionine, while histidine and arginine are produced by humans in such low amounts that dietary supplementation is recommended (especially histidine for children). Based on these data, quinoa certainly deserves its reputation as a source of high quality dietary protein.

REFERENCES

Brinegar, C., and Goundan, S., 1993, Isolation and characterization of chenopodin, the 11S seed storage protein of quinoa (*Chenopodium quinoa*), *J. Agric. Food Chem.* 41:182.

Brinegar, C., Sine, B., and Nwokocha, L., 1996, High-cysteine 2S seed storage proteins from quinoa (*Chenopodium quinoa*), *J. Agric. Food Chem.* 44: 1621.

Burnouf-Radosevich, M., 1988, Quinoa (*Chenopodium quinoa* Willd.): A potential new crop, in: *Biotechnology in Agriculture and Forestry, Vol. 6 Crops II*, Y. P. S. Bajaj, ed., Springer-Verlag, Berlin.

Burnouf-Radosevich, M., and Delfel, N. E., 1984, High-performance liquid chromatography of oleanane-type triterpenes, *J. Chromatogr.* 292:403.

Cardoza, A., and Tapia, M., 1979, Valor nutrivia, in: *Quinua y Kañiwa*, M. Tapia, ed., Serie Libros y Materiales Educativos No. 49, CIID-IICA, Bogotá.

Derbyshire, E., Wright, D. J., and Boulter, D., 1976, Legumin and vicilin, storage proteins of legume seeds, *Phytochem.* 15:3.

Fairbanks, D. J., Burgener, K. W., Robison, L. R., Andersen, W. R., and Ballon, E. R., 1990, Electrophoretic characterization of quinoa seed proteins, *Plant Breed.* 104:190.

Goundan, S., 1992, Isolation and characterization of chenopodin, the major seed storage protein of quinoa (*Chenopodium quinoa*), Master's Thesis, San Jose State Univ.

Johnson, D. C., and Croissant, R. L., 1989, Quinoa production in Colorado, *Serv. Action–Colo. State Univ. Coop. Ext.*, No. 112.

Mahoney, A. W., Lopez J. G., and Hendricks, D. G., 1975, An evaluation of the protein quality of quinoa, *J. Agric. Food Chem.* 23:190.

National Research Council, 1989, *Lost Crops of the Incas: Little-known Plants of the Andes with Promise for Worldwide Cultivation*, National Academy Press, Washington, D. C.

Rea, J., Tapia, M., and Mujica, A., 1979, Practicás agronómicas, in: *Quinua y Kañiwa*, Cultivos Andinos, M. Tapia, H. Gandarillas, S. Alandia, A. Cardozo, and A. Mujica, eds., CIID-IICA, Bogotá.

Risi, J. C., and Galwey, N. W., 1984, The Chenopodium grains of the Andes: Inca crops for modern agriculture, *Adv. Appl. Biol.* 10:145.

Varriano-Marston, E., and DeFrancisco, A., 1984, Ultrastructure of quinoa fruit (*Chenopodium quinoa*), *Food Microstruct.* 3:165.

Wood, T. R., 1985, Tale of a food survivor, *East West J.* 4:63.

MOLECULAR MECHANISM OF COMPETITIVE ADSORPTION OF α_{s1}-CASEIN AND β-CASEIN AT LIQUID INTERFACES

Srinivasan Damodaran

Department of Food Science
University of Wisconsin-Madison
1605 Linden Drive
Madison, WI 53706

INTRODUCTION

Food proteins are generally mixtures of several protein components. Thus, the foaming and emulsifying properties of commercial food proteins, such as egg-white, soy protein isolate, caseins, whey proteins, etc., are dependent on relative rates of binding and affinity of the protein components. Changes in composition during protein isolation or intentional manipulation of the composition of a protein mixture may alter the functional properties of the protein. Therefore, knowledge of the molecular factors that affect competitive adsorption of proteins at interfaces may be very useful in preparing protein ingredients that exhibit optimal functional properties in food systems.

COMPETITIVE ADSORPTION IN BINARY PROTEIN SYSTEMS

To elucidate the influence of one protein on the adsorption of another protein in binary protein mixtures, Damodaran et al (1-4) studied the kinetics of competitive adsorption of proteins from four binary protein systems, viz., β-casein/lysozyme, lysozyme/BSA, BSA/β-casein, and α_{s1}-casein/b-casein, at the air-water interface using a surface radiotracer method. The rationale for selecting these binary systems was that they represented various combinations of random coil/globular, globular/globular, and random coil/random coil proteins, as well as negatively/negatively, negatively/positively charged proteins. For instance, β-Casein and α_{s1}-casein represent random-coil-type hydrophobic and negatively proteins; lysozyme represents a highly rigid hydrophilic globular protein with a net positive charge; and BSA represents a negatively charged protein with a molecular flexibility some where those of β-casein and lysozyme.   Thus, the β-casein/lysozyme pair would represent random/globular and negatively/positively charged protein binary system; the lysozyme/BSA pair would represent globular/globular and positively/negatively charged protein binary system; the BSA/β-casein would present a globular/random and negatively/negatively charged protein binary system; and the α_{s1}-casein/β-casein would represent a random/random and negatively/negatively charged protein binary system. Thus, a fundamental understanding of the adsorption behavior of these proteins in the binary systems should provide the roles of charge-charge interactions, structural flexibility/rigidity, and hydrophilicity/hydrophobicity factors on competitive adsorption of proteins at interfaces.

One of the major conclusions of these studies was that, in binary protein systems involving random-coil/globular and globular/globular proteins, the competitive adsorption did not follow a Langmuir-type adsorption mechanism, which states that the interfacial concentrations of two proteins A ánd B at any bulk protein ratio should be

$$\Gamma_A = \frac{K_A C_A}{(1 + K_A a_A C_A + K_B a_B C_B)} \qquad [1]$$

and

$$\Gamma_B = \frac{K_B C_B}{(1 + K_A a_A C_A + K_B a_B C_B)} \qquad [2]$$

where Γ_A and Γ_B are the surface concentrations of A and B, respectively; K_A and K_B are equilibrium constants; a_A and a_B are the average area occupied per molecule of A and B, respectively, at monolayer coverage in single protein systems; and C_A and C_B are concentrations of A and B in the bulk phase at equilibrium.

This Langmuir-type adsorption mechanism for a binary system is based on the assumption that the surface concentrations of A and B at equilibrium is related to their relative binding affinities to the interface. That is, the composition of the binary film at the air-water interface at equilibrium is thermodynamically-controlled. However, in binary protein systems involving random coil/globular or globular/globular proteins, such as β-casein/lysozyme, lysozyme/BSA, and BSA/β-casein binary systems, this was not found to be the case (1-3). In these systems, the interfacial composition of the mixed protein film was primarily determined by the rate of arrival of each protein at the interface, and the molecular area available at the interface at the time of arrival. That is, the interfacial protein composition was kinetically-controlled. The protein that arrives first at the interface adsorbs first and it is not displaced by the late arriving protein component, even when the affinity of the latter to the interface is greater than that of the former.

An· example of the above phenomenon in the case of BSA/lysozyme binary system is shown in Figure 1. It should be noted that in single protein systems, adsorption of lysozyme begins only after about 110 min and reaches an apparent equilibrium surface concentration of about 0.67 mg m^{-2}, whereas BSA adsorbs immediately after creation of a fresh air-water interface, and the surface concentration reaches an equilibrium value of 0.9 mg m^{-2} within about 120 min. The important point to note here is that the adsorption of BSA is almost over even before lysozyme begins to adsorb to the interface. In the 1:1 binary system, the equilibrium surface concentration of lysozyme is only about 0.06 mg m^{-2}, whereas that of BSA is about 0.82 mg m^{-2}. In other words, in the 1:1 binary system, BSA completely suppresses adsorption of lysozyme to the air-water interface. This is not primarily because BSA has higher affinity than lysozyme to the interface, but because BSA arrives first at the interface, occupies most of the interfacial area before lysozyme arrives at the interface. The late arriving lysozyme does not displace the adsorbed BSA molecules from the interface. This is also true for BSA; that is, if lysozyme adsorbs first to the air/water interface, BSA cannot displace lysozyme from the interface.

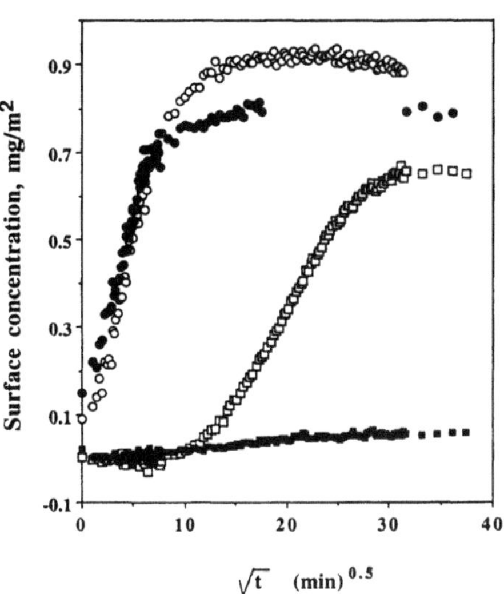

Figure 1: Time course of adsorption of lysozyme (squares) and bovine serum albumin (circles) at the air/water interface in single protein (open symbols) and in 1:1 binary mixture (filled symbols) systems. (From Ref. 2 with permission)

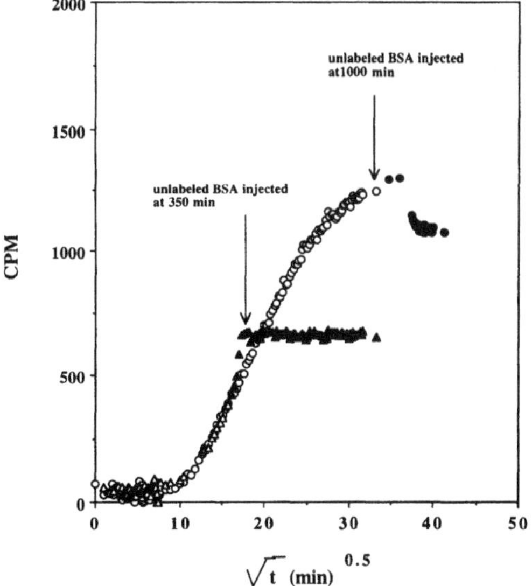

Figure 2: Displacement of adsorbed [14] C-lysozyme by unlabeled BSA. Symbols O and ● represent surface c.p.m of [14] C-lysozyme before and after injection, respectively, of unlabeled BSA at 1000 min adsorption. Symbols Δ and ▲ represent the same when unlabled BSA was injected at 350 min of adsorption. (From Ref. 2 with permission).

The above fact is further confirmed by the data shown in Figure 2 (2). In this case, ^{14}C-labeled lysozyme was first allowed to adsorb to the air-water interface; when unlabeled BSA was injected into the bulk phase at 350 min during the growth phase of adsorption of lysozyme, it abruptly stopped further adsorption of lysozyme to the interface. However, more interestingly, no desorption of the already adsorbed lysozyme occurred. Similarly, when unlabeled BSA was injected into the bulk phase after lysozyme has reached equilibrium adsorption (1000 min), only a small amount of lysozyme was desorbed by BSA from the interface. These results clearly indicate that once a globular protein is adsorbed to the interface, it cannot be displaced by another globular protein. Xu and Damodaran (3) have shown that this behavior is also true for random coil/globular protein binary systems. That is, a random coil-type protein cannot displace a globular protein from the air/water interface, and *vice versa*.

α_{s1}-Casein/β-Casein Binary System

However, a random coil/random coil protein binary system, for example the α_{s1}-casein/β-casein binary system, does not follow the above dictum. Figure 3 shows the kinetics of adsorption of α_{s1}-casein and β-casein to the air-water interface in single and binary protein systems. In single protein systems, both α_{s1}-casein and β-casein begin to adsorb immediately after a fresh air-water interface is created. The equilibrium surface concentration reaches a value of 1.66 mg m^{-2} for α_{s1}-casein and about 1.8 mg m^{-2} for b-casein. However, in the 1:1 binary system, the kinetics were more complex: The surface concentration of α_{s1}-casein increases first to a value of 1.0 mg m^{-2} within about 100 min and then decreases with time and reaches an equilibrium value of about 0.6 mg m^{-2}. In contrast, the surface concentration of β-casein increases continuously and reaches an equilibrium value of about 1.1 mg m^{-2}. The ratio of equilibrium surface concentration of α_{s1}-casein to β-casein is about 1:2. It should be noted that the time at which the sum of the surface concentrations of α_{s1}- and β-caseins (Γ_{total}) reaches a steady-state value coincides with the time at which Γ of α_{s1}-casein reaches its maximum value.

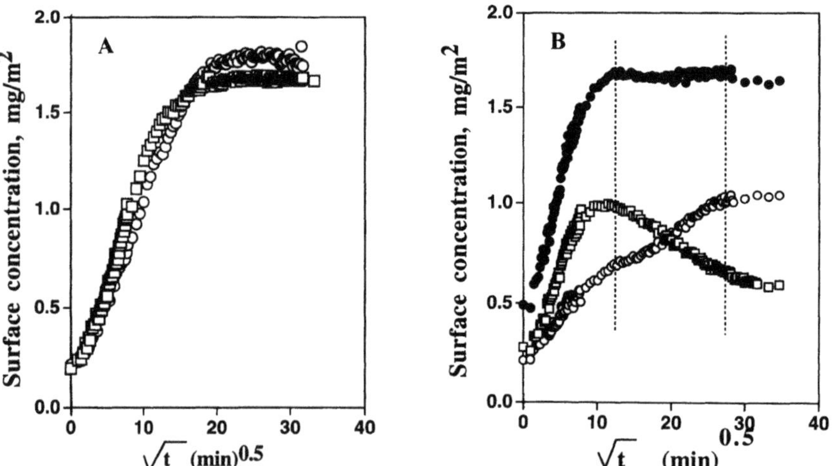

Figure 3: Time course of adsorption of α_{s1}-casein ($\square$) and β-casein ($\bigcirc$) at the air/water interface in (A) single-protein systems and (B) 1:1 binary mixture system. $\bullet$ represents total surface concentration of α_{s1}-casein plus β-casein as function of adsorption time. The vertical dotted lines denote the time zone in which displacement of α_{s1}-casein by β-casein occurs. (From Ref. 4 with permission).

Beyond this value, even though the surface concentration of α_{s1}-casein decreases and that of β-casein increases with time, the Γ_{total} remains unchanged. Since the molecular weights of α_{s1}-casein and β-casein are very close, this can be true only when one molecule of α_{s1}-casein is displaced from the interface for each β-casein molecule adsorbed to the interface. This in fact seems to be the case, because the rate of desorption of α_{s1}-casein, calculated from the slope of Figure 3 in the time zone indicated by the dotted lines, is almost the same as the rate of adsorption of β-casein in the same time zone.

To determine if α_{s1}-casein also can displace β-casein from the air/water interface, sequential adsorption of α_{s1}-casein and β-casein was studied (4). In this approach, first ^{14}C-labeled β-casein was allowed to adsorb to the air/water interface for 24 h. After 24 h, an aliquot of unlabeled α_{s1}-casein was injected into the bulk phase an the surface radioactivity of the adsorbed ^{14}C-β-casein was monitored as a function of time. If the unlabeled α_{s1}-casein displaced the adsorbed ^{14}C-β-casein from the air/water interface, then this should be reflected in a gradual decrease of surface radioactivity. Figure 4 shows the results of such studies (4). It should be noted that both α_{s1}-casein and β-casein could displace each other from the interface. This suggests that during adsorption from the bulk phase to the air/water interface, both α_{s1}-casein and β-casein must be continuously adsorbing and displacing each other at the interface. However, because β-casein is more surface active than α_{s1}-casein, it effectively displaces α_{s1}-casein from the interface. If we assume that both caseins are highly flexible proteins and that they experience no conformational constraints to unfold/spread at the air/water interface, then the differences in surface activities of these proteins must be related to differences in their mean residue hydrophobicity. The mean residue hydrophobicity of β-casein is 1330 cal/mol and that of α_{s1}-casein is about 1170 cal/mol. The small difference in mean residue hydrophobicity of these caseins seems to be sufficient enough to cause a large difference in their interfacial activity.

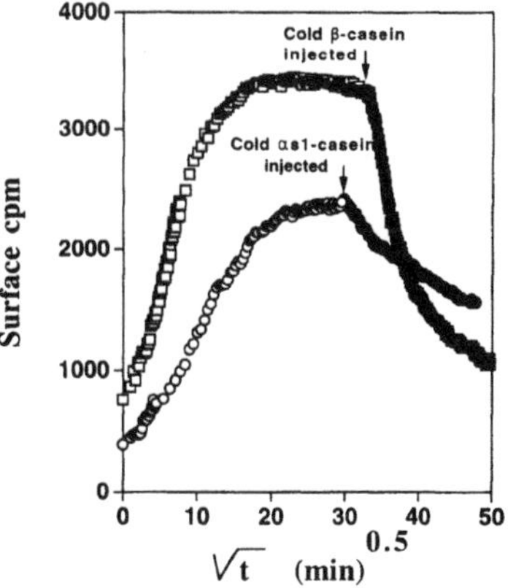

Figure 4: Displacement of ^{14}C-α_{s1}-casein by bulk phase unlabeled β-casein ($\square$, $\blacksquare$) and diplacement of ^{14}C-β-casein by bulk phase unlabeled α_{s1}-casein ($\bigcirc$). (From Ref. 4 with permission).

Role of Electrostatic Interactions in Adsorption. One of the notable features of the data in Figure 3 is that while initial rates of adsorption of α_{s1}-casein and β-casein in single protein systems are very similar, in the binary system the rate of adsorption of β-casein is slower than that of α_{s1}-casein. This fundamental change in the adsorption behavior of β-casein is at least partly due to the high negative charge of α_{s1}-casein in the mixed film. At any given surface concentration during the course of adsorption, the net charge density per unit area of an α_{s1}-casein plus β-casein mixed film is greater than that of β-casein alone film. Thus, the strong electrostatic repulsion between the partly formed α_{s1}-casein plus β-casein film at the interface and the approaching β-casein molecule from the bulk phase slows down the rate of adsorption of β-casein in the binary system.

The net charge of native β-casein is -13 and that of α_{s1}-casein is -21 (5). When all the eight serinephosphate residues of α_{s1}-casein are dephosphorylated, the net charge decreases to -9. To determine the role of net negative charge of α_{s1}-casein and β-casein on competitive adsorption and displacement at the air/water interface, the kinetics of adsorption of β-casein and dephosphorylated α_{s1}-casein in a 1:1 binary mixture was studied. Figure 5 shows that, the rate of adsorption of β-casein in the dephosphorylated α_{s1}-casein/β-casein binary system is faster than in the native α_{s1}-casein/β-casein binary system. It should also be noted that the surface concentration of dephosphorylated α_{s1}-casein increases first up to 0.9 mg m^{-2}, and then decreases with time to a final value of about 0.65 mg m^{-2}. On the other hand, the surface concentration of β-casein initially increases at the same rate as that of dephosphorylated α_{s1}-casein; however, when the concentration of dephosphorylated α_{s1}-casein reaches the maximum and begins its desorption phase (the second slope), β-casein continues to adsorb although at a slower rate. It should be also noted that the time at which Γ_{total} reaches a steady-state value coincides with the time at which Γ of dephosphorylated α_{s1}-casein reaches its maximum value and the rate of adsorption of β-casein slows down. This behavior is very similar to that of the native α_{s1}-casein/β-casein binary system (Figure 3), suggesting that the second phase of the kinetics (in the time zone indicated by the dotted lines) involves displacement of dephosphorylated α_{s1}-casein by b-casein. It should be noted that the ratio of equilibrium

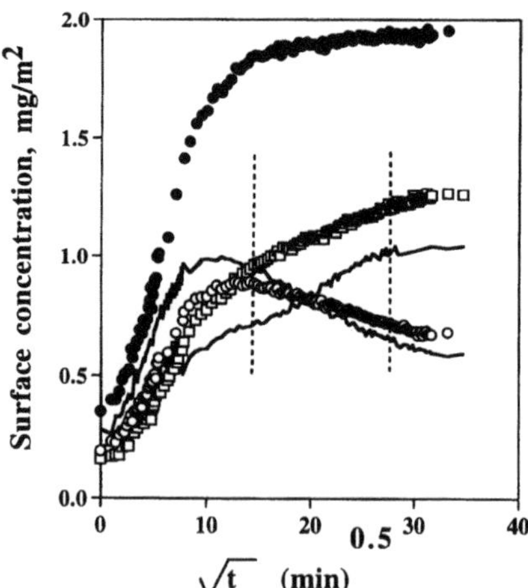

Figure 5: Time course of adsorption of dephosphorylated α_{s1}-casein (O) and β-casein (□) in a 1:1 binary mixture system. ● represents total surface concentration of dephosphorylated α_{s1}-casein plus β-casein as function of adsorption time. The vertical dotted lines denote the time zone in which displacement of dephosphorylated α_{s1}-casein by β-casein occurs. The solid lines are from Fig. 3, representing adsorption of native α_{s1}-casein and β-casein in a 1:1 binary system.

concentration of dephosphorylated α_{s1}-casein to β-casein is about 1:2, which is exactly same as in the case of native α_{s1}-casein/β-casein binary system. The data clearly show that although a decrease in the net negative charge of α_{s1}-casein by dephosphorylation slightly alters the rates of adsorption of α_{s1}-casein and β-casein, it does not change the overall dynamics of competitive adsorption and exchange of α_{s1}-casein and β-casein. The displacement of α_{s1}-casein from the air/water interface by β-casein occurs not because α_{s1}-casein has more net negative charge than β-casein, but because the hydrophobicity of β-casein is greater than that of α_{s1}-casein.

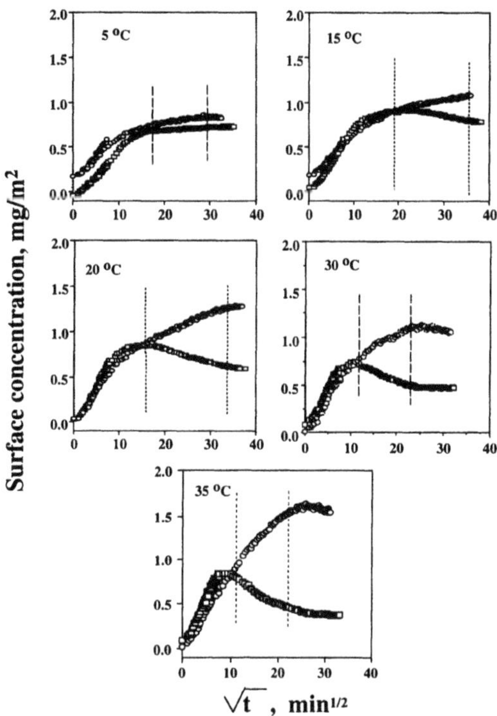

Figure 6: Effect of temperature on the kinetics of adsorption of α_{s1}-casein ($\square$) and β-casein ($\bigcirc$) at the air/water interface in a 1:1 binary system. The concentration of each protein in the binary mixture was 1.5 µg/ml.

Activation Energy Barrier for Adsorption and Displacement. Since displacement of α_{s1}-casein by β-casein occurs at constant Γ_{total} (Figure 3), the energy barrier for this displacement/adsorption process can be evaluated from the temperature-dependent changes in the rates of adsorption and desorption of β-casein and α_{s1}-casein, respectively, in the time zone indicated by the dotted lines in Figure 3. Figure 6 shows the kinetics of adsorption of α_{s1}-casein and β-casein from a 1:1 binary mixture at various temperatures. It should be noted that although the Γ_{total} was almost the same at all the temperatures studied, the rate of displacement of α_{s1}-casein by β-casein increased with temperature. The rate of adsorption of β-casein and the rate of displacement of α_{s1}-casein were determined from the slopes of the linear regions in the time zone indicated by the dotted lines. The Arrhenius plots for displacement of α_{s1}-casein and penetration of β-casein into the saturated monolayer of the α_{s1}-casein/β-casein mixed film are shown in Figure 7. The activation energy for penetration of β-casein into the mixed monolayer film is about 13.5 kcal/mol and that for displacement of α_{s1}-casein is about 10.5 kcal/mol. The difference of about 3 kcal/mol in the activation energies of these coupled processes is reasonable because, whereas desorption of α_{s1}-casein involves only the energy needed to detach bound segments from the air/water interface, adsorption of β-casein involves the energy needed to anchor at the interface against the surface pressure and to displace α_{s1}-casein from the interface. This tentatively suggests that the energy barrier for β-casein to anchor itself to the air/water interface against surface pressure of any protein film at the interface should be only about 3 kcal/mol.

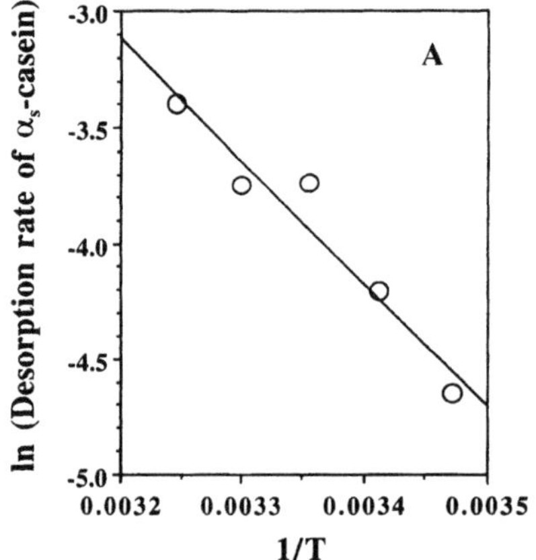
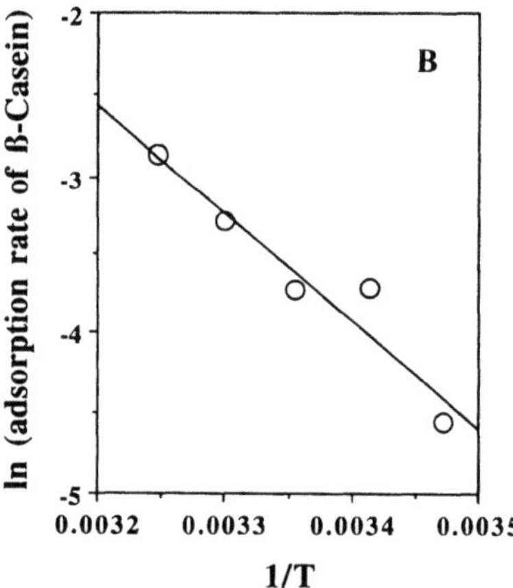

Figure 7: Arrhenius plots for desorption of α_{s1}-casein (A) and adsorption of β-casein in the time zone indicated by the dotted lines in Fig.6.

Thermodynamic compatibility. The ability of β-casein to displace α_{s1}-casein from the interface, and *vice versa*, can be arbitrarily attributed to the molecular flexibility of α_{s1}-

casein at the interface and therefore to its amenability to sequential detachment from the interface by β-casein. Similarly, the inability of β-casein to displace either lysozyme or BSA may be attributed *a priori* to relatively rigid conformation of these proteins at the interface and therefore the necessity to simultaneously detach all the adsorbed segments from the interface, which cannot be possible even by the highly surface active β-casein. Although such simplistic explanations are quite satisfactory, it is likely that a more fundamental mechanism, namely, thermodynamic compatibility, may be involved in these processes. In order for a protein to displace another protein from an interface, it should be able to mix (or dissolve) in the film of the latter. Since the volume-concentration of a protein film is generally very high (20-30%, w/v), dissolution of one protein into the interfacial film of another protein can occur only if the two proteins are thermodynamically compatible. Since α_{s1}-casein and β-casein are flexible disordered polymers belonging to a similar class of proteins, they are thermodynamically compatible with each other (6,7). This thermodynamic compatibility may allow β-casein to completely mix with α_{s1}-casein film (and *vice versa*) at the interface and thereby sequentially displace the segments of α_{s1}-casein from the interface purely on the basis of differences in affinities to the interface. However, lysozyme and β-casein belong to two different classes of proteins (i.e., globular *vs* disordered). They are thermodynamically incompatible and therefore they cannot mix in each other's films and

displace each other. It should be pointed out that bulk phase β-casein can displace or exchange with adsorbed β-casein (4), whereas bulk phase lysozyme cannot displace adsorbed lysozyme (2). This can be explained as follows: Since β-casein is a disordered protein, its conformation in the bulk phase and at the interface is very similar, that is, it goes from one disordered state to another isordercd state. Because of structural similarities, its physicochemical properties, notably the surface hydrophobicity/hydrophilicity character, remain the same and therefore they are still thermodynamically compatible. However, in the case of lysozyme, because of interfacial denaturation, the adsorbed lysozyme molecules are structurally different from those of the bulk phase molecules. Because of significant changes in the surface hydrophobic/hydrophilic character, probably the mixing of bulk phase lysozyme with the adsorbed (denatured) lysozyme in the film becomes thermodynamically incompatible. (Generally denatured proteins are thermodynamically incompatible with their native counterparts, which manifests itself in phase separation).

Thus, it appears that thermodynamic compatibility among proteins might be the fundamental mechanism by which displacement or exchange between proteins might occur at interfaces. This may also control the surface load and the composition of mixed protein films (1).

CONCLUSIONS

The results of these studies can be summarized as follows: A random-coil protein (such as α_{s1}-casein or β-casein) cannot displace an adsorbed globular protein, and a globular protein can neither displace nor exchange with an adsorbed globular protein nor displace a random-coil protein from the air-water interface. In contrast, a random coil protein can exchange or displace another random coil protein. In molecular terms what this means is that in the cases of globular/globular and globular/random-coil protein binary systems, adsorption essentially follows a noncompetitive (in a thermodynamic sense), irreversible mechanism; and in the case of random-coil/random-coil binary systems, adsorption follows a thermodynamically- controlled competitive, reversible mechanism. The fundamental factors that dictate the ability of a protein to displace another protein from the air/water interface are its thermodynamic compatibility of mixing with the latter and its hydrophobicity relative to the latter.

ACKNOWLEDGMENT

This work was support by a grant from the National Science Foundation (BCS 9315123).

REFERENCES

1. Y. Cao, and S. Damodaran, Coadsorption of β-casein and bovine serum albumin at the air-water interface from a binary mixture. *J. Agric. Food Chem.* 43, 2567-2573 (1995).
2. K. Anand, and S. Damodaran, Kinetics of adsorption of lysozyme and bovine serum albumin at the air-water interface from a binary mixture, *J. Colloid Interface Sci.* 176, 63-73 (1995).
3. S. Xu, and S. Damodaran, Kinetics of adsorption of proteins at the air-water interface from a binary mixture. *Langmuir* 10, 472-480 (1994).
4. K. Anand, and S. Damodaran, Dynamics of exchange between α_{s1}- casein and β-casein during adsorption at air-water interface. *J. Agric. Food Chem.* 44, 1022-1028 (1996).
5. H.E. Swaisgood, Chemistry of caseins, in *Advanced Dairy Chemistry*, P.F. Fox, ed., Elsevier Applied Science, Amsterdam (1992).
6. V.I. Polyakov, I.A. Popello, V.Y. Grinberg, and V.B. Tolstoguzov, Thermodynamic compatibility of proteins in aqueous medium. *Nahrung* 30, 365-368 (1986).
7. A. Prins, Principles of foam stability, in *Advances in Food Emulsions and Foams* (E. Dickinson, and G. Stainsby, eds.), Elsevier Applied Sci. Publishers, London and New York, 1988, pp. 91-122.

DISULFIDE-MEDIATED POLYMERIZATION OF WHEY PROTEINS IN WHEY PROTEIN ISOLATE-STABILIZED EMULSIONS

Frank J. Monahan[1], D. Julian McClements[2], and J. Bruce German[3]

[1]Department of Food Science
University College Dublin
Belfield, Dublin 4, Ireland

[2]Department of Food Science
University of Massachusetts
Amherst, MA 01003, U.S.A.

[3]Department of Food Science and Technology
University of California
Davis, CA 95616, U.S.A.

ABSTRACT

The effects of protein polymerization in whey protein isolate-stabilized emulsions on emulsion properties were investigated. Polymerization, involving intermolecular disulfide bonds between whey proteins adsorbed at the oil-water interface, increased with increasing storage time following emulsion formation. Ageing resulted in increased aggregation of emulsion droplets, emulsion viscosity and susceptibility to creaming but these effects were lower when thiol-disulfide interchange reactions were inhibited by N-ethylmaleimide (NEM). Following heating to 75°C, disulfide-mediated polymerization of whey proteins increased as did droplet aggregation, emulsion viscosity and creaming. While NEM lowered the extent of disulfide-mediated polymerization it did not affect the measured physical properties of the heated emulsions. Non-covalent interactions appeared to be the principal forces leading to aggregation of emulsion droplets but aggregates once formed were stabilized by disulfide bonds.

INTRODUCTION

Proteins containing the amino acid cysteine, either in the free thiol form or in the disulfide form, can undergo thiol oxidation and thiol-disulfide interchange reactions under certain conditions. These reactions result in the formation of new disulfide bonds (thiol oxidation) or rearrangement of existing ones (thiol-disulfide interchange) and have important consequences for protein structure and functionality in food systems. For example, disulfide bond formation occurs between the glutenins of wheat during bread dough manufacture (Ewart et al., 1968), between soy proteins in tofu manufacture (Saio et al., 1971) and between whey proteins in the formation of whey protein-stabilized emulsions (Dickinson and Matsumura, 1991).

In whey, the native β-lactoglobulin (β-Lg), α-lactalbumin (α-La), bovine serum albumin (BSA) and immunoglobulin (Ig) molecules are globular proteins with a tertiary structure stabilized by intramolecular disulfide bonds. β-Lg, α-La, BSA and Immunoglobulin G (IgG) have 5, 8, 35 and 64 cysteine residues, respectively (Kinsella et al., 1989). β-Lg and BSA each have a free thiol group.

β-Lg is the major protein component of whey and accounts for ~70% of the total protein (Morr and Foegeding, 1990). Between pH 2.0-3.7 and 5.1-8.0 β-Lg exists as a dimer and the free thiol group on each monomer is located in the interior of the globular protein and unavailable for interaction with free thiol groups or disulfide groups on other molecules. At pH 6.5 the dimer begins to dissociate and above pH 8.0 β-Lg exists as a globular monomer (Pessen et al., 1988). In this dissociated form the free thiol group is exposed and can engage in intra- or intermolecular thiol-disulfide interchange reactions (Dunnill and Greene, 1966). Unfolding of β-Lg induced by heat (Rector et al., 1989; Matsudomi et al., 1992), denaturants (Xiong and Kinsella, 1990) or by its use as an emulsifier (Dickinson and Matsumura, 1991) can also promote thiol-disulfide interchange.

The desirability of disulfide bond formation within or between molecules depends on the particular food application and the protein functionality sought. In whey protein gels, for example, intermolecular disulfide bonds increase gel firmness and elasticity (Mulvihill and Kinsella, 1988; Shimada and Cheftel, 1988). In foams and emulsions, extensively disulfide linked proteins which do not readily unfold, are less surface active than proteins with a more flexible structure (Kim and Kinsella, 1987; Kinsella and Phillips, 1989). The surface activity of such proteins can be improved by controlled reduction of disulfide bonds (Kim and Kinsella, 1987; Klemaszewski and Kinsella, 1991). Once emulsions form, however, surface film stability may be enhanced by disulfide bond formation (Dickinson, 1986).

This paper focuses on recent studies the authors have undertaken on intermolecular disulfide bond formation in whey protein isolate (WPI)-stabilized emulsions and the effects of disulfide-mediated polymerization of whey proteins on emulsion characteristics.

DISULFIDE-MEDIATED POLYMERIZATION IN WPI-STABILIZED EMULSIONS

In β-Lg and WPI-stabilized emulsions, whey proteins adsorbed at the oil-water interface undergo polymerization reactions involving the formation of disulfide bonds (Dickinson and Matsumura, 1991; McClements et al., 1993). In β-Lg-stabilized emulsions Dickinson and Matsumura (1991) showed, using SDS-PAGE with non-reducing conditions, that dimers, trimers and higher polymers of β-Lg were formed at the oil-water interface. Inclusion of the thiol blocking agent N-ethylmaleimide (NEM) in the β-Lg solution prior to emulsification inhibited polymerization providing further evidence for polymerization via disulfide bonds at the interface. The extent of disulfide-mediated polymerization increased with increasing time after emulsification. This suggested that unfolding of proteins at the interface occurred in a time-dependent manner, exposing free thiol groups and permitting increased intermolecular interaction and thiol-disulfide interchange with time.

WPI-stabilized emulsions demonstrated similar time-dependent interfacial polymerization (McClements et al., 1993; Monahan et al., 1993). The polymers formed were more complex because the four principal proteins present in whey - α-La, β-Lg, BSA and IgG - could all potentially engage in thiol-disulfide interchange reactions. In whey proteins adsorbed at the emulsion droplet interface, monomeric β-Lg and α-La disappeared with time and there was a concomitant appearance of high molecular weight polymers (Fig. 1) (Monahan et al., 1993). When free thiols were blocked with NEM polymerization was inhibited (Fig. 2) (McClements et al., 1993).

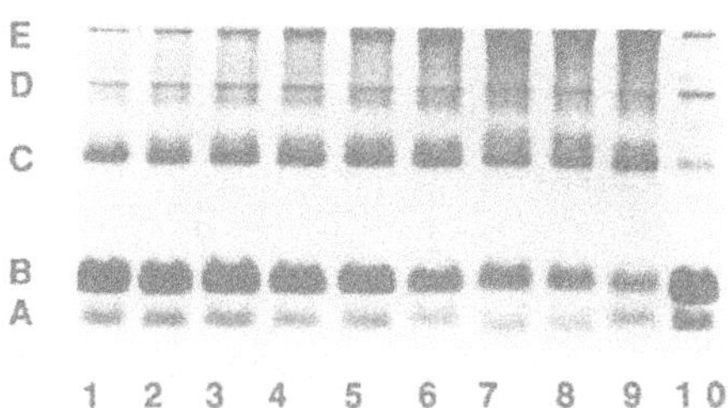

Figure 1. SDS-PAGE of adsorbed proteins from a 20 wt% n-hexadecane oil-in-water emulsion (stabilized with 1 wt% WPI) at various time intervals after emulsion formation. Lane 1: adsorbed protein immediately after emulsion formation. Lanes 2, 3, 4, 5, 6, 7, 8, 9: adsorbed protein obtained after the emulsions was stored at 20°C for 1, 2, 4, 6, 24, 48, 72 and 168 h, respectively. Lane 10: WPI solution used in emulsion preparation. (A) α-La (MW 14,200); (B) β-Lg (MW 18,400); (C) dimers of α-La and β-Lg; (D) BSA (MW 66,200); (E) Ig (MW 160,000). (from J. Agric Food Chem. 1993, 41, 1826-1829 with permission).

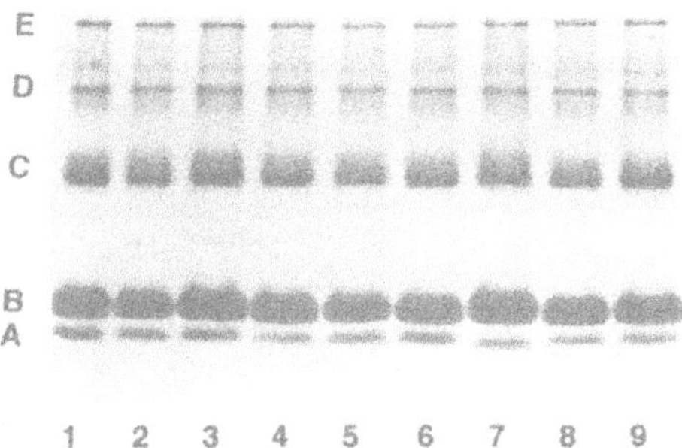

Figure 2. SDS-PAGE of adsorbed proteins from a 20 wt% n-hexadecane oil-in-water emulsion (stabilized with 1 wt% WPI) containing 0.5 mM NEM at various time intervals after emulsion formation. Lane 1: adsorbed protein immediately after emulsion formation. Lanes 2, 3, 4, 5, 6, 7, 8, 9: adsorbed protein obtained after the emulsions was stored at 20¡C for 1, 2, 4, 6, 24, 48, 72 and 168 h, respectively. (A) α-La (MW 14,200); (B) β-Lg (MW 18,400); (C) dimers of α-La and β-Lg; (D) BSA (MW 66,200); (E) Ig (MW 160,000). (from J. Food Sci. 1993, 58, 1036-1039, with permission).

Both α-La and β-Lg readily engaged in disulfide-mediated polymerization as demonstrated by the relationship between the disappearance of monomeric forms and the appearance of high molecular weight polymers with increasing time after emulsification (Monahan et al., 1993). When α-La alone was used as an emulsifier disulfide-mediated polymerization did not occur and this was attributed to its lack of a free thiol (Dickinson and

Matsumura, 1991). However, in emulsions containing both α-La and β-Lg the free thiol provided by β-Lg engages α-La in intermolecular disulfide bond formation (Dickinson and Matsumura, 1991). In WPI-stabilized emulsions a there was an initial increase in the formation of dimers of α-La and β-Lg followed by a decrease as the dimers engaged in further polymerization reactions to form higher molecular weight polymers (Fig. 1, Fig.3a,b,c) (Monahan et al., 1993).

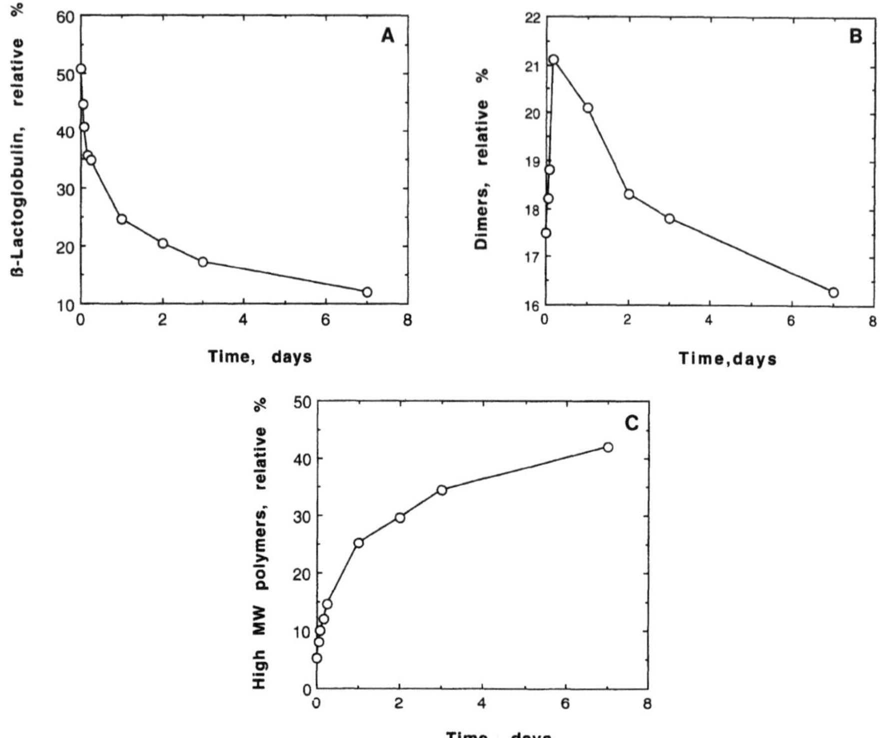

Figure 3. Effect of storage time at 20°C on the relative concentrations of (a) β-Lg, (b) dimers of β-Lg and α-La, and (c) high molecular weight polymers (MW > 160,000) in the adsorbed protein fraction of a 20 wt% n-hexadecane oil-in-water emulsion (stabilized with 1wt% WPI). (from J. Agric Food Chem. 1993, 41, 1826-1829 with permission).

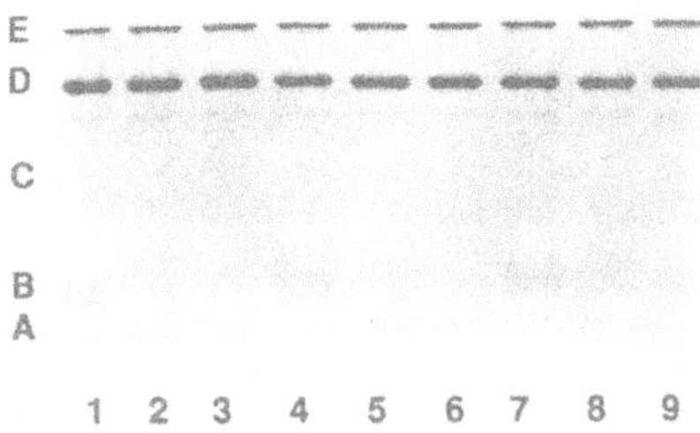

Figure 4. SDS-PAGE of unadsorbed proteins from a 20 wt% n-hexadecane oil-in-water emulsion (stabilized with 1wt% WPI) at various time intervals after emulsion formation. Lane 1: adsorbed protein immediately after emulsion formation. Lanes 2, 3, 4, 5, 6, 7, 8, 9: adsorbed protein obtained after the emulsions was stored at 20°C for 1, 2, 4, 6, 24, 48, 72 and 168 h, respectively. (A) α-La (MW 14,200); (B) β-Lg (MW 18,400); (C) dimers of α-La and β-Lg; (D) BSA (MW 66,200); (E) Ig (MW 160,000). (from J. Agric Food Chem. 1993, 41, 1826-1829 with permission).

130

An interesting observation in WPI-stabilized emulsions was that α-La and β-Lg were preferentially adsorbed over BSA and IgG at the oil-water interface (Monahan et al., 1993). When the electrophoretic profile of WPI in solution prior to emulsification (Fig 1, lane 10) was compared with that of adsorbed WPI immediately after emulsification (Fig 1, lane 1) the BSA and Ig bands in the latter were less intense. Furthermore, when the unadsorbed protein fraction was analysed it was predominantly composed of the BSA and Ig components (Fig. 4). Similarly, Shimizu et al. (1981) examined the adsorbed protein layer in a coconut oil-in-water emulsion stabilized with WPI and found that, at pH 7, β-Lg was selectively adsorbed and BSA was not adsorbed. In contrast to our study, they also found that both β-Lg and Ig were selectively adsorbed over both BSA and α-La at pH 7. However, the two studies are not directly comparable since the emulsions differed in oil type, buffer composition and protein concentration and these may account for the somewhat different findings.

TEMPERATURE EFFECTS ON DISULFIDE-MEDIATED POLYMERIZATION IN WPI-STABILIZED EMULSIONS

Treatments such as heating, pH alteration or addition of denaturants, which destabilize the structure of native whey proteins and promote unfolding, permit increased protein-protein interactions, including inter- and intramolecular thiol-disulfide interchange or thiol oxidation reactions (Matsudomi et al., 1992; Monahan et al., 1995; Xiong and Kinsella, 1990).

The combined effects of emulsification and post-emulsification heat treatment on disulfide-mediated polymerization was examined (Monahan et al., 1996). As expected, whey proteins adsorbed at the emulsion interface were extensively polymerized particularly after heating above 70°C (Fig. 5). In contrast to the unheated emulsions where NEM effectively inhibited time dependent polymerization, NEM reduced, but did not prevent, polymerization in heated emulsions (Fig. 5). This suggests that major structural changes brought about by heating make disulfide-mediated polymerization inevitable, even when a sulphydryl blocking agent is present.

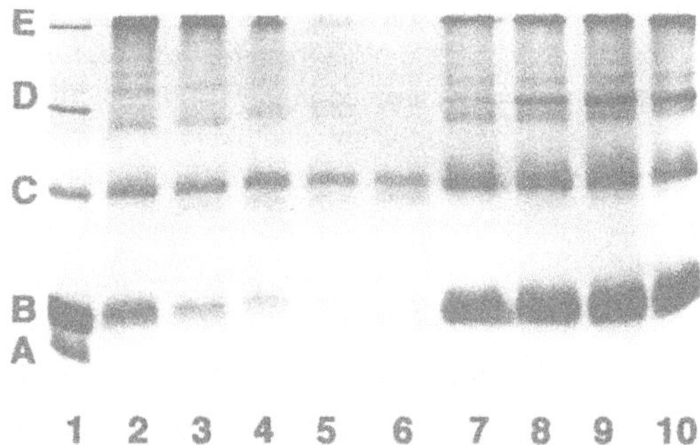

Figure 5. SDS-PAGE of a WPI solution in phosphate buffer (0.05 M, pH 7) (lane 1) and of adsorbed proteins from a 19% corn oil-in-water emulsion (stabilized by 1% WPI) heated at 30, 60, 70, 80, 90°C in the absence of NEM (lanes 2, 3, 4, 5 and 6, respectively) and heated at 30, 70, 80 and 90°C in the presence of NEM (lanes 7, 8, 9 and 10, respectively). A = α-La (MW, 14,200); B = β-Lg (MW, 18,400); C = dimers of α-La and β-Lg; D = BSA (MW, 66,200); E = Ig (MW, 160,000). (from J. Food Sci. 1996, 61, 504-509, with permission).

AGGREGATION OF EMULSION DROPLETS

Analysis of the particle size distribution of WPI-stabilized emulsions using laser light scattering showed that particle sizes increased with aging of emulsions (Fig. 6) (McClements et al., 1993). Increases in emulsion particle size can be attributed to the formation of aggregates of droplets in which the droplets either remain as distinct droplets (flocculate) or they coalesce into larger droplets. A method to distinguish between flocculated and coalesced droplets in a system in which an increase in droplet size is observed, involves displacing adsorbed interfacial proteins using a non-ionic surfactant such as Tween 20 (Courthaudon et al., 1991) and observing whether particle size is reduced, indicating that flocculates are present, or remains constant, indicating that droplets are coalesced. Addition of Tween 20 to aged emulsions led to a decrease in particle size detected by light scattering and indicated that the large particles formed following aging were in fact flocculates and not coalesced droplets (McClements et al., 1993).

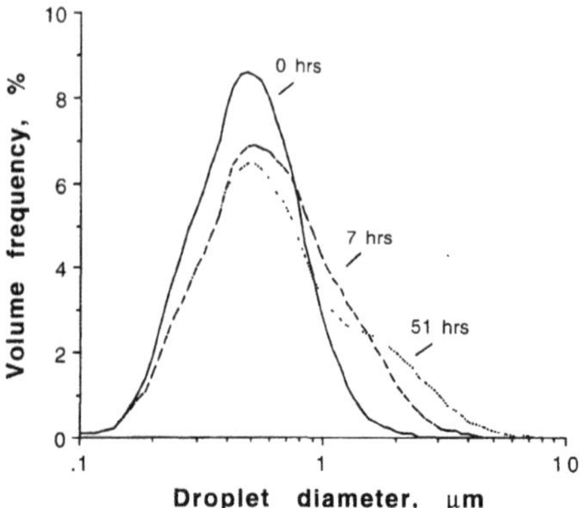

Figure 6. Effect of storage at 20°C on droplet size distributions of 20 wt% n-hexadecane oil-in-water emulsions (stabilized with 1 wt% WPI). (from J. Food Sci. 1993, 58, 1036-1039, with permission).

Disulfide bond formation was demonstrated to have a role in the aggregation process since inclusion of NEM in the emulsions, at a level calculated to block all free thiols, reduced the extent of aggregation (Fig. 7) (McClements et al., 1993). Aggregation of droplets with time was not prevented by inclusion of NEM indicating that other interdroplet interactions, such as hydrogen, ionic, hyydrophobic or van der Waals interactions, between adsorbed proteins on neighbouring droplets, were probably involved in aggregation. Similar to the thiol-disulfide interchange reactions, these other reactions are also likely to be enhanced by time-dependent unfolding of whey proteins at the emulsion droplet interface.

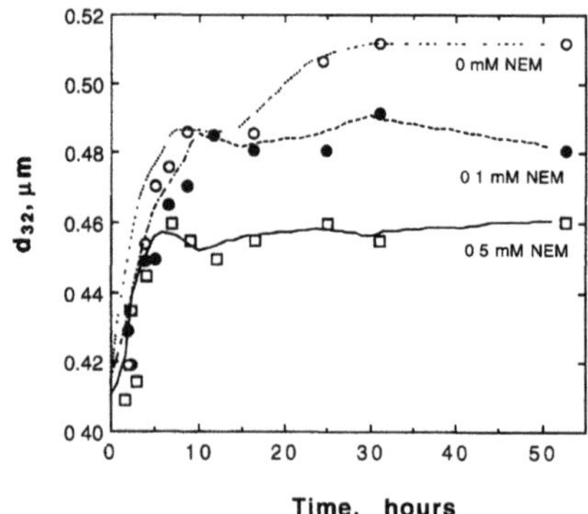

Figure 7. Changes in mean droplet diameter d_{32} with time after emulsion formation for samples from a 20 wt% n-hexadecane oil-in-water emulsion (stabilized with 1 wt% WPI) containing different concentrations of NEM. Higher NEM concentrations (up to 20 mM) gave results similar to those of 0.5 mM samples. (from J. Food Sci. 1993, 58, 1036-1039, with permission).

Heating WPI-stabilized emulsions above the denaturation temperatures of the whey proteins increased the aggregation of emulsion droplets (Monahan et al., 1996). As in the unheated system, SDS-PAGE indicated that disulfide-mediated polymerization occurred between proteins adsorbed at the emulsion interface (Figure 5). However, the thiol blocking agent did not affect aggregation, as observed with the unheated emulsions (Fig. 7). Nevertheless, the aggregates could be more easily dissociated by a non-ionic surfactant when NEM was present (Fig. 8). This suggests that non-covalent forces are likely to be involved in aggregation and that the aggregates may be stabilized by disulfide bonds.

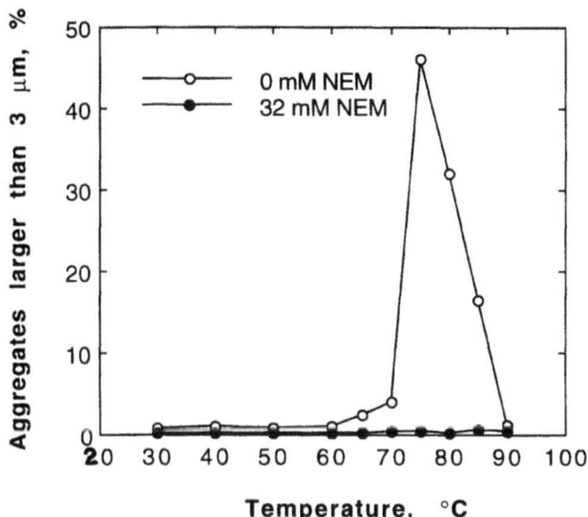

Figure 8. Effect of addition of Tween 20 (0.002%) and holding for 24 h on percentage of aggregates > 3 mm diameter in 19 wt% corn oil-in-water emulsions (stabilized by 1% WPI) containing 0 and 32 mM NEM, heated at various temperatures for 30 min, held for a minimum of 15 h at 25°C and diluted to an oil concentration of 0.04% with phosphate buffer. (from J. Food Sci. 1996, 61, with permission).

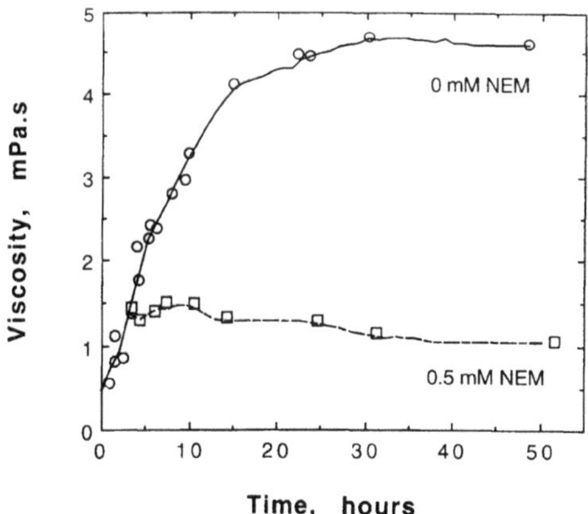

Figure 9. Changes in the viscosities of 20 wt% n-hexadecane oil-in-water emulsions (stabilized with 1wt% WPI) emulsions containing different concentrations of NEM with time after emulsion formation. (from J. Food Sci. 1993, 58, 1036-1039, with permission).

EMULSION VISCOSITY

In emulsions in which extensive aggregates formed and were stabilized by disulfide bonds apparent viscosity was higher than that of emulsions containing less extensive aggregates (Fig. 9) (McClements et al.,1993). In unheated emulsions preventing thiol-disulfide interchange by addition of NEM resulted in a decrease in emulsion viscosity (Fig. 9). The lower apparent viscosity of the emulsions in which disulfide-mediated polymerization was prevented was probably the result of less extensive aggregation in these emulsions.

In heated emulsions the apparent viscosity increased dramatically above 70°C irrespective of whether or not thiol groups were blocked (Fig. 10) (Monahan et al., 1995). The increase in apparent viscosity is probably related to the extent of aggregate formation (Fig. 7 and 10). The absence of a major effect due to the inclusion of NEM suggests that non-covalent forces enhance aggregation and are likely to be more important determinants of droplet aggregation and apparent viscosity than disulfide bonds.

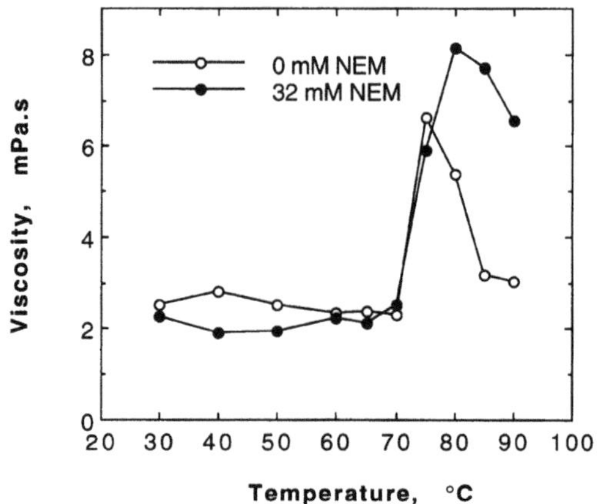

Figure 10. Apparent viscosity of 19 wt% corn oil-in-water emulsions (stabilized by 1% WPI), containing 0 and 32 mM NEM, heated at various temperatures and held for a minimum of 15 h at 25¡C. (from J. Food Sci. 1996, 61, 504-509, with permission).

EMULSION STABILITY

In unheated emulsions, blocking thiol groups with NEM, and thereby inhibiting thiol-disulfide interchange, led to a reduction in the rate of creaming of emulsions (McClements et al., 1993). The increased stability could result from the lower aggregation of droplets in emulsions in which thiol-disulfide interchanges are blocked. In heated emulsions, any effect on emulsion stability of blocking thiol-disulfide interchange with NEM was counteracted by the effect of other interactions also promoted by heating (Fig. 11) (Monahan et al., 1996). In both unheated and heated systems the susceptibility of emulsions to cream appears to be related to the degree of aggregation of droplets, with more extensive creaming in emulsions containing larger aggregates (Tornberg and Hermansson, 1977). A decrease in the tendency of emulsions to cream was observed after heating to 85-90°C compared to heating to 75-80°C. We suggested that at the higher temperatures intradroplet interactions may be greater than interdroplet (droplet-droplet) and that a more dense protein network forms around individual droplets. Droplets stabilized by a dense interfacial film can have an effective density larger than that of the dispersion medium and are therefore less susceptible to creaming (Dickinson and Stainsby, 1982).

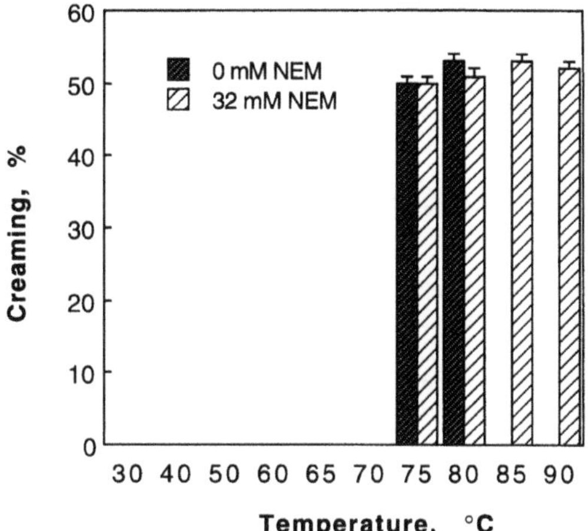

Figure 11. Effect of centrifugation (1000g, 30 min., 25°C) on the formation of a creamed layer in 19 wt% corn oil-in-water emulsions (stabilized by 1% WPI), containing 0 and 32 mM NEM, heated at various temperatures and held for a minimum of 15 h at 25°C. (from J. Food Sci. 1996, 61, 504-509, with permission).

CONCLUSION

The data presented suggest that disulfide-mediated polymerization of whey proteins my affect the properties of WPI-stabilized emulsions, particularly emulsion droplet aggregation, emulsion viscosity and stability to creaming. It is clear that disulfide bond formation is, however, not the sole factor involved in aggregation of emulsion droplets or in changes in emulsion viscosity or creaming. Rather, disulfide bond formation accompanies other non-covalent molecular interactions which may be of equal or greater significance in influencing emulsion characteristics. Similarly, in heated emulsions, while disulfide-mediated polymerization of whey proteins is greatly increased and emulsion characteristics significantly altered, non-covalent interactions are probably the major determinants of heat-induced changes in emulsion characteristics.

REFERENCES

Courthaudon, J.-L., Dickinson, E., Matsumura, Y., and Williams, A., 1991, Influence of emulsifier on the competitive adsorption of whey proteins in emulsions, *Food Structure* 10:109.

Dickinson, E., 1986, Mixed proteinaceous emulsifiers: review of competitive protein adsorption and the relationship to food colloid stabilization, *Food Hydrocolloids* 1:3.

Dickinson, E., and Matsumura, Y., 1991, Time-dependent polymerization of β-lactoglobulin through disulfide bonds at the oil-water interface in emulsions, *Int. J. Biol. Macromol.* 13:26.

Dickinson, E., and Stainsby, G., 1982, *Colloids in Foods*, Applied Science Publishers Ltd., London.

Dunnill, P., and Green, D.W., 1966, Sulphydryl groups and the N - R conformational change in β-lactoglobulin, *J. Molec. Biol.* 15:147.

Ewart, J.A.D., 1968, A hypothesis for the structure and rheology of glutenin. *J. Sci. Fd. Agric.* 19:617.

Kim, S.H., and Kinsella, J.E., 1987, Surface active properties of food proteins: effects of reduction of disulfide bonds on film properties and foam stability of glycinin, *J. Food Sci.* 52:128.

Kinsella, J.E., and Phillips, L.G., 1989, Structure:function relationships in food proteins, film and foaming behaviour, in: *Food Proteins*, Kinsella, J.E. and Soucie, W.G., eds., American Oil Chemists' Society, Champaign, IL.

Kinsella, J.E., Whitehead, D.M., Brady, J., and Bringe, N.A., 1989, Milk proteins: possible relationships of structure and function, in: *Developments in Food Chemistry 4. Functional Milk Proteins*; P.F. Fox, ed., Elsevier Applied Science Publishers, London.

Klemaszewski, J.L., and Kinsella, J.E., 1991, Sulfitolysis of whey proteins: effects on emulsion properties, *J. Agric Food Chem.* 39:1033.

Matsudomi, N., Oshita, T., Sasaki, E., and Kobayashi, K., 1992, Enhanced heat-induced gelation of β-lactoglobulin by α-lactalbumin, *Biosci. Biotech. Biochem.* 56:1697.

McClements, D.J., Monahan, F.J., and Kinsella, J.E., 1993, Disulfide bond formation affects the stability of whey protein isolate emulsions, *J. Food Sci.* 58:1036.

Monahan, F.J., McClements, D.J., and German, J.B., 1996, Effect of heat on disulfide-mediated polymerization reactions and physical properties of WPI-stabilized emulsions, *J. Food Sci.* 61, 504-509.

Monahan, F.J., German, J.B., and Kinsella, J.E., 1995, Effect of pH and temperature on protein unfolding and thiol/disulphide interchange reactions during heat-induced gelation of whey proteins, *J. Agric. Food Chem.* 43:46.

Monahan, F.J., McClements, D.J., and Kinsella, J.E., 1993, Polymerization of whey protein in whey protein-stabilized emulsions, *J. Agric. Food Chem.* 41:1826.

Morr, C.V., and Foegeding, E.A., 1990, Composition and functionality of commercial whey and milk protein concentrates and isolates: a status report, *Food Technol.* 44(4):100.

Mulvihill, D.M., and Kinsella, J.E., 1988, Gelation of β-lactoglobulin: effects of sodium chloride and calcium chloride on the rheological and structural properties of gels, *J. Food Sci.* 53:231.

Pessen, H., Kumosinski, T.F., and Farrell, H.M., 1988, Investigation of differences in the tertiary structures of food proteins by small angle x-ray scattering, *J. Ind. Microbiol.* 3:98.

Rector, D.J., Kella, N.K., and Kinsella, J.E., 1989, Reversible gelation of whey proteins: melting, thermodynamics and viscoelastic behaviour, *J. Text. Studies* 20:457.

Saio, K., Kajikawa, M., and Watanabe, T., 1971, Food processing characteristics of soybean proteins. Part II. Effect of sulfhydryl groups on physical properties of tofu-gel, *Agric. Biol. Chem.* 35:890.

Shimada, K., and Cheftel, J.C., 1988, Texture characteristics, protein solubility, and sulfhydryl group/disulfide bond contents of heat-induced gels of whey protein isolate, *J. Agric. Food Chem.* 36:1018.

Shimizu, M., Kamiya, T., and Yamauchi, K., 1981, The adsorption of whey proteins on the surface of emulsified fat, *Agric. Biol. Chem.* 45:2491.

Tornberg, E., and Hermansson, A.-M., 1977, Functional characterization of protein stabilized emulsions: effect of processing, *J. Food Sci.* 42:468.

Xiong, Y.L., and Kinsella, J.E., 1990, Mechanism of urea-induced whey protein gelation, *J. Agric. Food Chem.* 38:1887.

PARTIAL COALESCENCE AND STRUCTURE FORMATION IN DAIRY EMULSIONS

H. Douglas Goff

Associate Professor
Department of Food Science
University of Guelph
Guelph, Ontario, N1G 2W1
Canada

INTRODUCTION

The process of controlled partial coalescence of an emulsion during whipping and air incorporation is responsible for the establishment of structure in two notable dairy products, whipped cream and ice cream, leading to complex products described both as protein-stabilized emulsions and fat-stabilized foams. This process has been studied by several researchers in whipped cream (Schmidt and van Hooydonk, 1980; Walstra and Jenness, 1984; Brooker et al., 1986; Noda and Shiinoki, 1986; Anderson et al., 1987; Bruhn and Bruhn, 1988; Brooker, 1990; Needs and Huitson, 1991; Stanley et al., 1996) and ice cream (Govin and Leeder, 1971; Lin and Leeder, 1974; Goff and Jordan, 1989; Berger, 1990; Gelin et al., 1994; Kokubo et al., 1994). Partial coalescence is also responsible for structure formation in a variety of whipped non-dairy dessert toppings (Buchheim et al., 1985; Barfod and Krog, 1987; Krog et al., 1987). In 1988, Prof. John Kinsella co-supervised a Ph.D. dissertation prepared by myself on the topic of protein-emulsifier interactions at the fat globule interface and its relation to partial coalescence and structure formation in ice cream (Goff, 1988). This paper briefly reviews the topic of partial coalescence and describes further work completed by our research group at the University of Guelph since the completion of that dissertation. Prof. Kinsella is thus responsible for stimulating interest in this subject, which is an ongoing part of active research at present.

Emulsions are inherently unstable. The interfacial tension or surface free energy between fat and water surfaces is high. Rearrangements will lower the surface free energy as a thermodynamic system moves toward equilibrium by reducing the surface area. The equilibrium situation would be minimum free energy, minimum surface area, and two distinct layers, fat and water (Walstra, 1987). In addition, density differences between the fat and water phases result in a driving force for the fat phase to rise, a process known as creaming (Walstra and Jenness, 1984). Therefore, creaming and coalescence are inevitable in emulsions and both lead to loss of the dispersed state. However, amphiphilic molecules play an important role in emulsions. Proteins with hydrophobic regions or mono- and di-glycerides are examples of such molecules as they contain segments that prefer solution into an aqueous environment and segments that prefer solution into a non-polar environment (Aynié et al., 1992; Tomas et al., 1994). Thus, during the homogenization of a fat into a solution in the presence of amphiphilic molecules, a membrane quickly forms around the fat globule. This membrane acts to lower the oil water interfacial tension and, depending on the amount of surfactant adsorbed, increase the density of the fat globule (Walstra, 1987). Both have a stabilizing effect, slowing down the rate of creaming and coalescence that may have otherwise occured.

PARTIAL COALESCENCE

In an oil-in-water emulsion, when the membrane between two approaching globules is ruptured, the fat generally flows together, causing the loss of identity of the original particle and creating a larger one. Many of the triglyceride emulsions common in the food industry, particularly dairy emulsions, are stored at temperatures where fat crystallization occurs (Goff et al., 1987; Boode et al., 1991). Such globules show intricate patterns of crystals within the globule (Fig. 1a), and crystals actually growing or protruding through the membrane (Fig. 1b). Crystals can be found both radial and tangential to the surface, indicating that a solid shell, liquid core model is not appropriate to describe partially crystalline globules (Boode and Walstra, 1993). If crystals are present in the oil phase, coalescence may be incomplete, leading to the formation of irregularly aggregated globules that retain some of their original identity, but are intricately linked (Fig. 2). This process has been referred to as partial coalescence. This form of aggregation exhibits some important differences compared to true coalescence of liquid globules: the identity of the individual globules is still retained in the aggregates; due to the irregular form of the aggregates, the viscosity of the emulsion may increase; the aggregation can proceed until a continuous network is formed throughout the volume, thus giving the product solid properties (yield stress) and immobilizing other

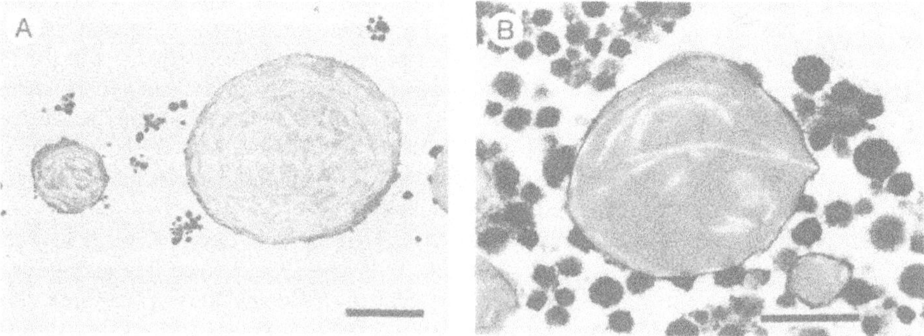

Figure 1. A) Milkfat globule showing the intricate structure of needle-like crystals present in the cross-section after fixation at 4°C and examination by TEM. B) Globule distorted by the presence of the internal crystalline structure. (Bar=0.5 µm). From Goff et al., 1987.

Figure 2. A schematic representation of the network of fat globules formed during partial coalescence, showing the important role of fat crystals in cementing the coalescing globules together (visualized as the straight lines within the globule. Courtesy of Prof. P. Walstra, Wageningen, Netherlands.

particles (eg., air cells) present; the rate of aggregation greatly depends on agitation, while liquid droplets rarely show an appreciable dependence of coalescence on agitation; during flow or agitation, the stability to partial coalescence is orders of magnitude smaller than would be the case for true coalescence, ie., the globules containing no crystals (Darling 1982; Boode, 1992; Boode and Walstra, 1993; Boode et al., 1993).

Aggregation can occur via perikinetic (brownian motion) or orthokinetic (shear-induced) mechanisms. A very important feature of partial coalescence is the role that both agitation and air incorporation have on its onset, ie., the large differences in rate between perikinetic and orthokinetic aggregation. van Boekel and Walstra (1981a,b) demonstrated that velocity gradients in liquid emulsions containing partially-crystalline fat globules increased the rate of partial coalescence by a factor of 10^6. The most probable explanation for this shear-dependence is that crystals protruding from the O/W interface pierce the thin aqueous film between closely approaching globules (Darling, 1982; Boode and Walstra, 1993). Air has also been shown to greatly affect the formation of fat clustering in partially crystalline emulsions, both when the emulsion is in the static state (eg., sparging) and during agitation (Walstra, 1987). Thus, the combination of air and agitation produces extremely rapid partial coalescence, as can easily be demonstrated by whipping a bowl of heavy cream. This process will be discussed in detail further on.

Another important factor to consider is the role of the membrane in the stability of the emulsion to partial coalescence. Williams and Dickinson (1995) showed that orthokinetic stability of a β-lactoglobulin emulsion increased as the concentration of protein used in the manufacture of the emulsion increased, due both to smaller average droplet sizes with increased protein, and also to the amount of protein adsorbed at the interface (per μm^2, the surface excess), which increases as the amount of protein used is increased. Britten et al. (1994) showed that resistance to stirring-induced coalescence was improved by increasing the proportion of denatured whey protein isolate when compared to native WPI. Increasing proportions of denatured WPI increased the emulsifying activity of the protein blends, increased emulsion viscosity, and increased emulsion stability. Proteins adsorb to surfaces by orienting their hydrophobic regions toward the fat interface. If protein unfolding has occurred prior to emulsion formation, more of the hydrophobic regions should be exposed, such that the extent of denaturation was correlated with the formation of more intact, stronger membranes. Important differences have been shown in the behaviour of various milk proteins to partial coalescence (Goff et al., 1989), with whey protein emulsions generally being more susceptible than casein emulsions.

STRUCTURE FORMATION

Ice Cream

Ice cream is a complex food colloid that consists of air bubbles, fat globules, ice crystals, and an unfrozen serum phase. Ice crystals and air bubbles are usually in the range of 20-50 μm and 50-100 μm, respectively. The air bubbles are usually lined with fat globules and the fat globules are coated with a protein/ emulsifier layer (Caldwell et al., 1992). The serum phase consists of the sugars and high molecular weight polysaccharides in a freeze-concentrated solution. Various steps in the manufacturing process, including pasteurization, homogenization, aging, freezing, and hardening, contribute to the development of this structure. To properly describe the role of partial coalescence in structure formation, it is necessary to begin with the formation of the emulsion at the time of homogenization and the role of the ingredients present at the time of homogenization, with particular reference to the fat, proteins, and emulsifiers. Following homogenization, the emulsion is further affected by changes occuring during the aging step, viz., crystallization of the fat and rearrangement of the fat globule membrane to the lowest free energy state. This emulsion then undergoes both whipping and ice crystallization during the dynamic freezing process, which contributes to the development of the three main structural components of the frozen product: a discontinuous foam, a network of partially coalesced fat surrounding the air bubbles, and ice crystals. Since the major focus of the research that was done with Prof. Kinsella on this topic focussed on the role of the small-molecule surfactants in this process, that is the subject which will be described in detail here.

Homogenization begins the process of fat structure formation. After preheating or pasteurization, the mix is at a temperature sufficient to have melted all the fat present, and the fat passes through one or two homogenizing valves. Immediately following homogenization, the newly formed fat globule is practically devoid of any membranous material and readily adsorbs amphiphilic molecules from solution (Darling and Butcher, 1978; Walstra and Oortwijn, 1982). The immediate environment supplies the surfactant molecules, which include caseins, undenatured whey proteins, phospholipids, lipoprotein molecules, components of the original milkfat globule membrane, and any added chemical surfactants (Oortwijn and Walstra, 1979, Berger, 1990). The membrane formed during homogenization continues to develop during the aging step and rearrangement occurs until the lowest possible energy state is reached (Barfod et al., 1991, Berger, 1990). The transit time through an homogenization valve is in the order of 10^{-5} to 10^{-6} seconds (Pandolfe, 1982). Protein unfolding, adsorption and rearrangement at the interface may take minutes or even hours to be complete. It is clear, therefore, that the immediate membrane formed upon homogenization is a function of the microenvironment at the time of its creation, and that the recombined membrane of the fat globule in the aged mix is not fully developed until well into the aging process (Berger, 1990).

Emulsifiers are not needed in an ice cream mix to stabilize the fat emulsion (Table 1), due to an excess of protein and other amphiphilic molecules in solution (Govin and Leeder, 1971; Goff and Jordan, 1989). If a mix is homogenized without any emulsifier, both the whey proteins and the caseins will form this new fat globule membrane, with the caseins contributing much more to the bulk of the adsorbed protein (Figure 3a). However, if surfactants such as polysorbate 80 or mono- and di-glycerides are present, they have the ability to lower the interfacial tension between the fat and the water phases lower than the proteins. It has been reported by us that the adsorption of milk proteins to fat globules lowered the interfacial tension of the fat/serum interface from 8.26 dynes/cm to 5.5 dynes/cm. However, the addition of a surfactant (polysorbate 80) to the system lowered the interfacial tension further than was accomplished by the proteins alone, to 2.24 dynes/cm in the presence of the milk protein (Goff and Jordan, 1989). Thus they become preferentially adsorbed to the surface of the fat (Darling, 1982; Walstra and Jenness, 1984; Goff and Jordan, 1989; Barfod et al., 1991). As the interfacial tension is lowered and proteins are displaced from the surface of the fat, the quantity of adsorbed material is reduced, eg., from 15.9% of the total protein in the ice cream mix to 7.8% in the presence of the polysorbate 80 (Table 2), and the actual membrane becomes weaker to subsequent destabilization. This is due to the fact that the protein molecules, and particularly the caseins, are considerably larger than the emulsifier molecules such that a membrane made up entirely of emulsifier is very thin (Figure 3b), i.e., lower surface excess, although the emulsion is thermodynamically favoured due to the lowering of the interfacial tension and net free energy of the system. The emulsifier had no effect on the size distribution of the globules in the mix, as determined by TEM micrographs. However, Dickinson et al. (1993) point out that subsequent instability to shear is not induced by the protein displacement *per se* , but by some change in the adsorbed layer properties caused by the replacement of a pure protein monolayer by a mixed protein/surfactant monolayer.

The next stage of structure development occurs during the concomitant whipping and freezing step. Air is incorporated either through a lengthy whipping process (batch freezers) or drawn into the mix by vacuum or injected under pressure (continuous freezers) (Berger, 1990). This process causes the emulsion to undergo partial coalescence or fat destabilization, during which clumps and clusters of the fat globules form and build an internal fat structure or network into the frozen product, in a very analogous manner to the whipping of heavy cream (Darling, 1982; Brooker et al., 1986; Berger, 1990). Cross-linking of fat globules from one air cell to the next, thus forming an infrastructure as may be the case in whipped cream, is less likely in ice cream due to the reduction in dispersed phase volume from the heavy cream system to the ice cream mix system, however, it must also be borne in mind that the air bubbles, fat globules, and aqueous phase are being freeze-concentrated at the same time.

The fat globule clusters formed during the process of partial coalescence are responsible for surrounding and stabilizing the air cells and creating a semi- continuous network or matrix of fat throughout the product resulting in the beneficial properties of dryness upon extrusion during the manufacturing stages (aids in packaging and novelty molding, for example), a smooth-eating texture in the frozen dessert, and resistance to meltdown or good

Table 1. Comparison of the creaming rates of ice cream mix emulsions either with no surfactant or with 0.08% Tween 80 when both have sat quiescently for 10 days. Butterfat determined by Mojonnier ether extraction. From Goff et al., 1987.

	No Emulsifier	Polysorbate 80
Butterfat (%) in mix	10.00	10.00
Butterfat (%), Top of 100ml Cylinder, 10 days	11.25 ± 0.43	11.28 ± 0.43
Butterfat (%), Bottom of 100ml Cylinder, 10 days	9.47 ± 0.11	9.46 ± 0.11

n = 4

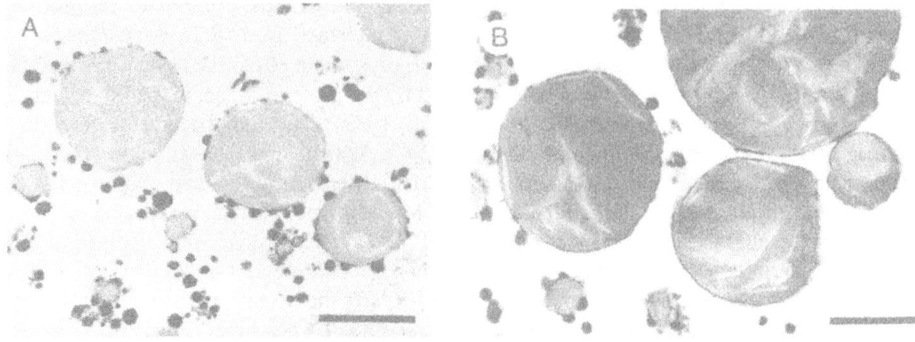

Figure 3. Transmission electron micrograph of an ice cream mix emulsion in the A) absence and B) presence of 0.08% polysorbate 80. Bar=1 μm. From Goff et al., 1987.

stand-up properties (necessary for soft serve operations) (Berger, 1990; Lin and Leeder, 1974; Buchheim et al., 1985). It has long been recognized that fat destabilization is enhanced by some of the emulsifiers in common use (Figure 4). When the emulsion is subjected to the tremendous shear forces in the barrel freezer, the thin membrane created by the addition of surfactant is not sufficient to prevent the fat globules from coalescing during collision, thus setting up the internal fat matrix. If an ice cream mix is subjected to excessive shearing action or contains too much emulsifier, the formation of objectionable butter particles can occur as the emulsion is churned beyond the optimum level. It should be possible to predict the destabilizing power of an emulsifier in the barrel freezer by measuring the interfacial tension between the fat and water phases in the presence of the emulsifier to determine the extent of protein coverage that will result and by considering the molecular size (and surface excess) of

Table 2. Differences in the amount of protein adsorbed to the fat globules in each of the mixes as determined by centrifugal separation and Kjeldahl analyses. From Goff et al., 1987.

	No Emulsifier	Polysorbate 80 (0.08%)
Mix Protein (%)	4.08 ± 0.01	4.08 ± 0.01
Protein in serum		
(% of mix)	3.43 ± 0.07	3.76 ± 0.17
Adsorbed Protein		
(% of total protein)	15.9	7.84

n = 4

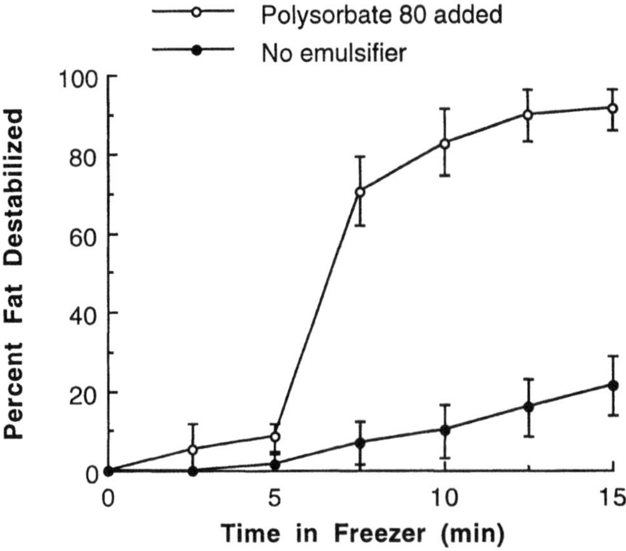

Figure 4. The effect of emulsifier addition on the percent fat destabilized (determined by turbidity of a 1:500 dilution at 5°C) during the period of time the ice cream mix was held in the barrel of a batch freezer. Ice crystallization began after about 5 minutes (Goff et al., 1988).

the emulsifier. Polysorbate 80, having a small molecular weight and producing the lowest interfacial tension, displaces more protein, resulting in a very thin membrane, and thus produces the maximum amount of fat destabilization (Goff et. al., 1987).

Changes to the fat emulsion have recently been studied by Gelin et al. (1994). They have demonstrated through light scattering measurements of fat globule size distribution and aggregation that the freezing step is responsible for considerable fat aggregation, but that this aggregation is reversible through dissociation with SDS. It was obvious from their study that the homogenization step accounted for a large amount of adsorbed protein, and that casein was preferentially adsorbed over the whey protein. The aging and freezing-hardening-thawing steps each accounted for subsequent protein desorption, again mostly of the caseins. The emulsifier used in these mixes contained saturated mono- and di-glycerides but no polysorbate 80, which, based on the results of Goff and Jordan (1989), may have led to greater amounts of protein desorption during the aging period. Kokubo et al. (1994) have also demonstrated, in a study of the relationship between draw temperature and overrun of ice cream and resulting de-emulsified fat, that increased fat de-emulsification as a result of decreased draw temperature or increased overrun caused changes in the fat particle size distribution. Particles ≤ 1.2 μm decreased while particles 3-4 μm and 8-15 μm increased in frequency with increased fat de-emulsification, as a result of fat globule clustering. The sequential process of partial coalescence during ice cream freezing has also been examined (Goff and Jordan, 1989). The incorporation of air alone, or the shearing action alone, independent of freezing, are not sufficient to cause the same degree of fat destabilization as when ice crystallization occurs concomitantly. The freezing process causes an increase in concentration of the mix components, such as proteins and mineral salts, in the unfrozen water phase. It is believed that the ice crystals contribute to the shearing action on the fat globules, due to their physical shape, and that the concentration of components also leads to enhanced destabilization. However, to create the desired fat destabilization, whipping and freezing must occur simultaneously (Goff, 1988).

Whipped Cream

Whipped cream is another dairy product that relies heavily on partial coalescence for the development of structure, as it is converted from a viscous liquid into a viscoelastic solid during the process of whipping. Figure 5 illustrates the build-up of the semi-continuous network of fat surrounding and stabilizing the air bubbles. In studying the whipping of heavy cream, Schmidt and vanHooydonk (1980) stated that the proteinaceous membrane that envelops the air bubble is penetrated by fat globules as the whipping process proceeds, and this fat penetration offers foam stability to the whipped product. Brooker et al. (1986) offered a more detailed explanation of the whipping and foam stabilization process in heavy cream. During the initial stages of whipping, air bubbles were stabilized primarily by β-casein and whey proteins with little involvement of fat. Adsorption of fat to air bubbles occurred when the fat globule membrane coalesced with the air water interface. Only rarely did fat spread at the air water interface. The final cream was stabilized by a cross linking of fat globules surrounding each air cell to adjacent air cells thus building an infrastructure in the foam. In skim milk foams, the initial air water interface is also formed by the serum proteins and soluble β-casein with little involvement of micellar casein. Micelles became attached as a discontinuous layer but were not deformed or spread (Brooker, 1985). Thus, the important role of the protein, both in the serum phase and at the membrane, cannot be overlooked in the development of whipped cream structure. Bruhn and Bruhn (1988) observed that UHT cream took about 40% longer to whip than raw or pasteurized cream. Recalling the results of Britten et al. (1994) related to the stability of emulsions with denatured WPI, it is easily postulated that the UHT treated cream resulted in greater amounts of protein adsorption and/or increased membrane integrity, ultimately increasing the stability to partial coalescence.

We have recently studied the effect of the protein membrane composition on partial coalescence in 20% milkfat emulsions. Emulsions were produced consisting of 20% fat

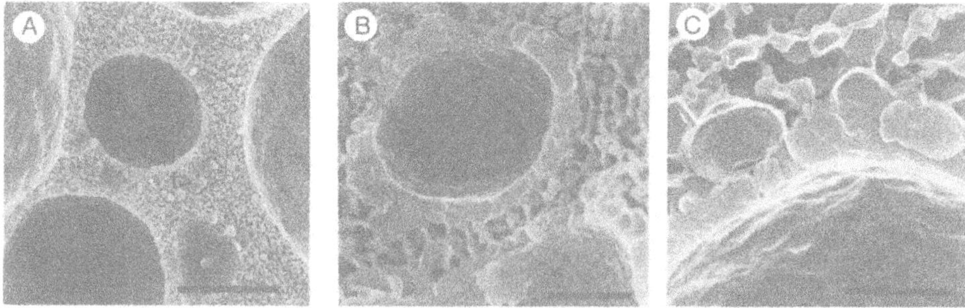

Figure 5. The structure of whipped cream as determined by scanning electron microscopy. A) Overview showing the relative size and prevalence of air bubbles (a) and fat globules (f); bar = 30 μm. B) Internal structure of the air bubble, showing the layer of partially coalesced fat which has stabilized the bubble; bar = 5 μm. C) Details of the partially-coalesced fat layer, showing the interaction of the individual fat globules; bar = 3 μm.

(from sweet butter, 80% fat), and 0.085, 0.17, 0.25, 0.50, 0.75, 1.0, 1.5 or 2.0% whey protein isolate (WPI, 88% protein, Protose Separations, Teeswater, ON). The emulsions were heated to 70° or 90°C for 30 min, continuously stirred with a hand mixer, and then homogenized at 3000 psig, 1000 psig on the second stage (Cherry Burrell Superhomo A125A). Emulsion stability was determined by measuring the fat depletion at the bottom of a 100 ml graduated cylinder in 48 hr at 4°C by Babcock fat analysis. Emulsions were whipped with a Sunbeam mixmaster and aliquots removed each min for overrun measurement, fat destabilization by spectroturbidity and particle size analysis (Malvern Mastersizer X). Foam stability was assessed by removing aliquots after 2 min of whipping and measuring collapse and drainage after 24, 48, and 72 hr.

As protein concentration was increased from 0.085% to 0.75%, the mean fat globule diameter decreased, levelling off at a $d_{3,2}$ of 0.6 μm at protein concentrations greater than 0.75%. Emulsions with greater than 0.25% protein were found to be stable to creaming under quiescent conditions. However, during whipping and agitation, considerable differences were seen in the behaviour of the emulsions (Fig. 6). Less than 10% fat destabilization was seen with protein concentration of greater than 1%. However, with protein concentrations between 0.25 and 1%, substantial destabilization was noted during the time course of whipping. Between 0.25% and 0.5% protein, the resulting foams showed little collapse or drainage over 72 hours. When the fat globule size distribution in these emulsions was examined by a Malvern Mastersizer X, it was evident that a small number of larger globules/clusters had formed (Fig. 7), which were responsible for the resulting foam stiffness. It was thus concluded that a range of protein concentrations existed where the initial size of the fat globule was small, the emulsion was stable to creaming in the quiescent state, the emulsion formed significant fat destabilization during whipping and agitation, and the resulting foam was stable to collapse and drainage. This is very significant, given that the fat content in these emulsions was only 20%.

The above research indicates the potential for the development of lower fat whipping creams for the retail market. Preliminary work has been conducted by us to test the concept

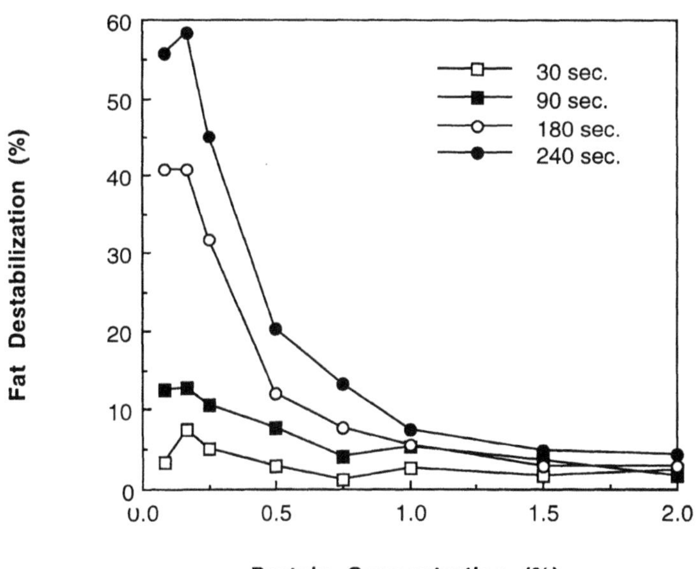

Figure 6. Fat destabilization, determined by spectroturbidity, in 20% milkfat emulsions as a function of protein concentration in the emulsion and time of whipping in a kitchen mixer at high speed.

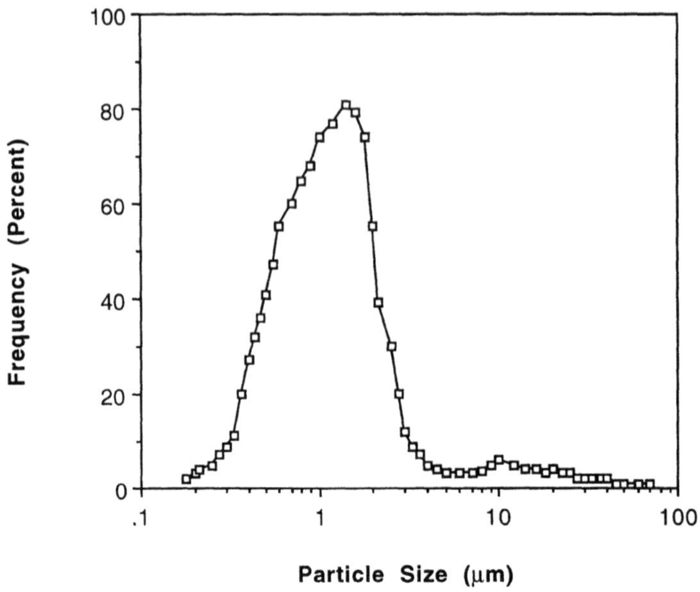

Figure 7. Fat globule size distribution of a partially-coalesced emulsion after whipping, demonstrating the formation of a small number of large globules/clusters

of altering the fat globule membrane composition to affect whipped cream formation and stability. Emulsions were prepared containing 18% w/w butterfat from sweet butter, 0.175 - 0.55% whey protein isolate (92% protein), 4.0 - 6.0% msnf (skim milk powder, 97% solids), 5.0 - 8.0% sugar, and 0.45 - 0.60% stabilizer. The emulsions were prepared in two stages by heating the butterfat, WPI and a portion of the water to 70°C for 30 min, homogenizing at 3000 psig, 1000 psig on the second stage, cooling to 4°C, then blending with the skim milk powder, sugar, stabilizer and remainder of the water, which had previously been heated to 70°C for 30 minutes and cooled to 4°C. Emulsion stability was tested by examination of creaming after 24 hrs. in a 100 ml graduated cylinder. Overrun and foam stability were examined by whipping the emulsions with a Sunbeam home mixer for 2 min. Several combinations of WPI, sugar, and stabilizer produced emulsions that were stable to creaming before whipping and produced stable foams after whipping. None, however, matched the control (35% fat real cream) for foam stiffness or consumer mouthfeel. While this project was only a very preliminary examination of product development applications, it suggests that application of a two step process of membrane formation and post-homogenization blending with other desired ingredients, particularly milk proteins not wanted at the fat interface, produced a stable emulsion capable of being whipped into a stable foam.

CONCLUSIONS

Emulsion stability and protein surface layers have been well studied in model systems. Partial coalescence has also been studied in model systems to gain a good understanding of the mechanisms involved, but to a much lesser extent than the study of emulsions. Partial coalescence has been demonstrated as the causative agent in structure formation in two notable dairy products, viz., ice cream and whipped cream. The challenge that lies ahead for the dairy scientist is to apply the knowledge of the model system research to the creation of new and/or modified dairy products. An example of such application is in the development of the low fat dairy-based whipping cream system described above. While this development is a long way from a commercial reality, it demonstrates that utilization of the concepts of surface excess differences between whey proteins and caseins and the process of selective homogenization to control the membrane-forming species can produce emulsions that demonstrate stable foams during whipping, which is a manifestation of optimal partial coalescence.

ACKNOWLEDGEMENTS

Appreciation is expressed to the Ontario Ministry of Agriculture, Food and Rural Affairs for financial support, and to Shanti Anant and Sandy Smith for collaboration in the whipped cream studies.

REFERENCES

Anderson, M., B. E. Brooker and E. C. Needs. 1987. The role of proteins in the stabilization/destabilization of dairy foams, in: Food Emulsions and Foams, E. Dickinson, ed. Royal Society of Chemistry, London, pg. 100.

Aynié, S., M. Le Meste, B. Colas and D. Lorient. 1992. Interactions between lipids and milk proteins in emulsion. J. Food Sci. 57: 883.

Barfod, N. M. and N. Krog. 1987. Destabilization and fat crystallization of whippable emulsions (toppings) studied by pulsed NMR. J. Amer. Oil Chem. Soc. 64: 112.

Barfod, N. M., N. Krog, G. Larsen and W. Buchheim. 1991. Effect of emulsifiers on protein-fat interaction in ice cream mix during aging I. Quantitative Analyses. Fat Sci. Technol. 93: 24.

Berger, K. G. 1990. Ice cream, in: Food Emulsions, 2nd edn, K. Larsson and S. Friberg, eds. Marcel Dekker Inc., New York, pg. 367.

Boode, K. 1992. Partial coalescence in oil-in:-water emulsions. Ph.D. Dissertation, Wageningen Agricultural University, Netherlands.

Boode, K., C. Bisperink and P. Walstra. 1991. Destabilization of O/W emulsions containing fat crystals by temperature cycling. Colloids and Surfaces 61: 55.

Boode, K. and P. Walstra. 1993. Partial coalescence in oil-in-water emulsions 1. Nature of the aggregation. Colloids and Surfaces A: Physicochem. and Eng. Aspects 81: 121.

Boode, K., P. Walstra and A. E. A. deGroot-Mostert. 1993. Partial coalescence in oil-in-water emulsions 2. Influence of the properties of the fat. Colloids and Surfaces A: Physicochem. and Eng. Aspects 81: 139.

Britten, M. H. J. Giroux, Y. Jean, and n. Rodrigue. 1994. Composite blends from heat denatured and undenatured whey protein: emulsifying properties. Int. Dairy J. 4: 25.

Brooker, B. E. 1985. Observations on the air serum interface of milk foams. Food Microstruc. 4: 289.

Brooker, B. E. 1990. The adsorption of crystalline fat to the air-water interface of whipped cream. Food Struc. 9: 223.

Brooker, B. E., M. Anderson, and A. T. Andrews. 1986. The development of structure in whipped cream, Food Microstruc. 5: 277.

Bruhn, C. M. and J. C. Bruhn. 1988. Observations on the whipping characteristics of cream. J. Dairy Sci. 71: 857.

Buchheim, W., N. M. Barfod, and N. Krog. 1985. Relation between microstructure, destabilization phenomena and rheological properties of whippable emulsions. Food Microstruc. 4: 221.

Caldwell, K. B., H. D Goff, and D. W. Stanley. 1992. A low - temperature scanning electron microscopy study of ice cream. I. Techniques and general microstructure. Food Structure. 11: 1-9 .

Darling, D. F. 1982. Recent advances in the destabilization of dairy emulsions. J. Dairy Res. 49: 695.

Darling, D. F. and D.W. Butcher. 1978. Milkfat globule membrane in homogenized cream, J. Dairy Res. 45: 197.

Dickinson, E., R. K. Owusu, and A. Williams. 1993. Orthokinetic destabilization of a protein-stabilized emulsion by a water-soluble surfactant, J. Chem Soc. Faraday Trans.89: 865.

Gelin, J.-L., L. Poyen, J. L. Courthadon, M. Le Meste and D. Lorient. 1994. Structural changes in oil-in-water emulsions during the manufacture of ice cream, Food Hydrocoll. 8: 299.

Goff, H. D. 1988. The role of chemical emulsifiers and dairy proteins in fat destabilization during the manufacture of ice cream. Ph.D. Dissertation, Cornell Univ., Ithaca, NY.

Goff, H.D. and W.K. Jordan. 1989. Action of emulsifiers in promoting fat destabilization during the manufacture of ice cream. J. Dairy Sci. 72: 18 - 29.

Goff, H.D., J.E. Kinsella, and W.K. Jordan. 1989. Influence of various milk protein isolates on ice cream emulsion stability. J. Dairy Sci. 72: 385 - 397.

Goff, H.D., M. Liboff, W.K. Jordan, and J.E. Kinsella. 1987. The effects of Polysorbate 80 on the fat emulsion in ice cream mix : evidence from transmission electron microscopy studies. Food Microstructure. 6 : 193 - 198.

Govin, R. and J. G. Leeder. 1971. Action of emulsifiers in ice cream utilizing the HLB concept. J. Food Sci. 36: 718.

Kokubo, S., K. Sakurai, M. Hattori, and M. Tomita. 1994. Effect of drawing temperature at a freezer and overrun on deemulsified fat of ice cream. J. Japanese Soc. Food Sci. Technol. 41: 347. (Dairy Sci. Abst., 1995, 57: 3397).

Krog, N., N. M. Barfod, and W. Buchheim. 1987. Protein-fat-surfactant interactions in whippable emulsions, in: Food Emulsions and Foams, E. Dickinson, ed. Royal Society of Chemistry, London, pg. 144.

Lin, P. M. and J. G. Leeder. 1974. Mechanism of emulsifier action in an ice cream system. J. Food Sci. 39: 108.

Melsen, J. P. and P. Walstra. 1989. Stability of recombined milk fat globules. Neth. Milk Dairy J. 43: 63.

Needs, H. C. and A. Huitson. 1991. The contribution of milk serum proteins to the development of whipped cream structure. Food Struc. 10: 353.

Noda, M. and Y. Shiinoki. 1986. Microstructure and rheological behaviour of whipping cream. J. Texture Studies. 17: 189.

Oortwijn, H., and P. Walstra. 1979. The membranes of recombined fat globules. 2. Composition. Neth. Milk Dairy J. 33: 134.

Pandolfe, W. D. 1982. Development of the New Gaulin Micro-Gap Homogenizing Valve, J. Dairy Sci. 65: 2035.

Schmidt, D. G. and A.C.M. vanHooydonk. 1980. A scanning electron microscopal investigation of the whipping of cream. Scanning Electron Microscopy III: 653.

Stanley, D. W., H. D. Goff and A. S. Smith. 1996. Texture-structure relationships in foamed dairy emulsions. Food Research Internat. In press.

Tomas, A. , J.-L. Courthadon, D. Paquet, and D. Lorient. 1994. Effect of surfactant on some physico-chemical properties of dairy oil-in-water emulsions. Food Hydrocoll. 8: 543.

van Boekel, M. A. J. S. and P. Walstra. 1981. Effect of couette flow on stability of oil-in-water emulsions. Colloids and Surfaces 3: 99.

van Boekel, M. A. J. S. and P. Walstra. 1981. Stability of oil-in:-water emulsions with crystals in the disperse phase. Colloids and Surfaces 3: 109.

Walstra, P. 1987. Overview of emulsion and foam stability, in: Food Emulsions and Foams, E. Dickinson, ed. Royal Society of Chem, London, pg. 242.

Walstra, P., and H. Oortwijn. 1982. The membranes of recombined fat globules. 3. Mode of formation. Neth. Milk Dairy J. 36: 103.

Walstra, P., and R. Jenness. 1984. 'Dairy Chemistry and Physics'. John Wiley and Sons, New York.

Williams, A. and E. Dickinson. 1995. Shear-induced instability of oil-in-water emulsions, in: Food Macromolecules and Colloids, E. Dickinson and D. Lorient, eds., Royal Soc. Chem., London, pg. 252.

SOLUBILIZATION OF OIL DROPLETS BY MICELLAR
SURFACTANT SOLUTIONS

D. Julian McClements

Biopolymers and Colloids Laboratory
Department of Food Science
University of Massachusetts
Amherst, MA 01003

INTRODUCTION

Traditionally, small molecule surfactants have been used in the food industry for three major purposes: (i) to enhance the formation and stability of emulsions; (ii) to modify food structure and rheology by interaction with biopolymers; and (iii) to alter the morphology of fat crystals by interaction with triacylglycerols (Krog 1990, Dickinson 1992). Recently, a number of novel applications of surfactant molecules have been identified which are likely to lead to the development of new processing techniques in the near future (Dickinson and McClements 1995, Wolcott 1995). These applications are based on the ability of small molecule surfactants to form molecular aggregates called *micelles* (Figure 1). Typically, a micelle consists of between 30 and 100 surfactant molecules, orientated so that their hydrophobic tails are located in the interior of the micelle (away from the water), and their hydrophilic head groups are located at the exterior (in contact with the water). This type of structure is formed because it minimizes the unfavourable contact area between polar and non-polar regions. In contrast to emulsions, micelles are thermodynamically stable systems. They also have highly dynamic and flexible structures, with surfactant molecules rapidly entering and leaving the micelle. Even so, they do have a clearly defined average size (distribution) and shape over measurable time scales. Typically, micelles are between 5 and 20 nm in diameter, and are therefore about a thousand times smaller than emulsion droplets.

One of the most important properties of micelles is their ability to incorporate non-polar molecules into their hydrophobic interiors (Hiemenz 1986, Dickinson and McClements 1995). Non-polar molecules can either be solubilized between the hydrocarbon tails of the surfactant molecules or form a separate "pool" in the interior of the micelle (or both), depending on their concentration and molecular structure. Once solubilized in micelles non-polar molecules can be transported across an aqueous phase in which they are normally insoluble (McClements *et al* 1992, McClements and Dungan 1993). The ability of surfactant micelles to solubilize and transport oil molecules has a number of important

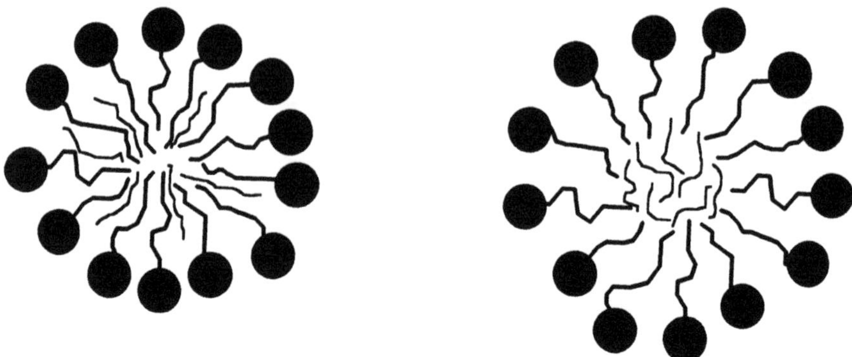

Figure 1. Surfactant micelles can solubilize non-polar molecules in their hydrophobic interiors.

implications and possible applications in the food industry:

- The distribution of ingredients in foods depends on their partitioning between oil, aqueous and interfacial regions, which is influenced by the presence of surfactant micelles (Wedzicha 1988).
- Micelles can be used to incorporate non-polar molecules, such as vitamins, flavors, essential oils or antioxidants, into an aqueous solution (Han *et al.* 1990, El-Nokaly *et al* 1991, Wolcott 1995).
- The ability of surfactant micelles to incorporate and transport oil molecules offers a useful method for introducing lipophilic ingredients into emulsion droplets, through an aqueous phase in which they are normally insoluble. The lipophilic ingredient is injected into an emulsion in the form of swollen-micelles or emulsion droplets, and is redistributed throughout the system because of oil exchange between micelles and droplets (McClements and Dungan 1993, Dickinson and McClements 1995). This technique could be used to incorporate oil-soluble vitamins, colors, antioxidants or flavors into pre-existing emulsions.
- The addition of surfactant micelles to emulsions may prove to be a useful means of controlling flavor release (McNulty and Karel 1973, McNulty 1987).
- Surfactant micelles may also be used to selectively extract specific non-polar components (*e.g.* cholesterol) from emulsified or bulk oils (Wolcott 1995).
- Micelles can also be used to control the rate of certain chemical reactions (Hiemenz 1986).

Studies of the factors which affect the solubilization of non-polar molecules in surfactant molecules are therefore extremely beneficial to the food industry, because they may lead to the development of new processing techniques based on micellar technology. Previous studies have shown that the rate of oil solubilization depends on the surface area of oil exposed to the aqueous phase, the type of emulsifier and oil used, the temperature and the presence of co-surfactants (McBain 1955, Nakagawa 1966, Elworthy *et al* 1968, Carroll 1981, Mackay 1987, Ward 1989, Aveyard *et al* 1990, Ward and Quigley 1990). Most of these studies have been been carried out using bulk oils or relatively large oil droplets in contact with an aqueous micellar solution. Recently, the author and co-workers have developed two novel methods, one based on differential scanning calorimetry (McClements *et al* 1992, McClements and Dungan 1993) and the other on laser light scattering (McClements and Dungan 1996, Weiss *et al* 1996a,b, Coupland *et al* 1996), to study the kinetics of emulsion droplet solubilization in surfactant micelles. These techniques have enabled us to quantify factors affecting the solubilization of emulsion droplets, as well as giving us a greater insight into the physical mechanisms by which solubilization occurs. This chapter describes recent experimental work carried out at the University of Massachusetts to elucidate the factors which affect the rate and extent of oil solubilization of emulsion droplets in surfactant micelles.

EXPERIMENTAL PROCEDURES

Materials

Polyoxyethylene sorbitan monolaurate (Tween 20), *n*-tetradecene, *n*-tetradecane, *n*-hexadecene, *n*-hexadecane, *n*-octadecene and *n*-octadecane were obtained from the Sigma Chemical Company (St. Louis, MO). Corn oil was purchased from a local supermarket. Double distilled and deionized water was used in the preparation of all solutions and emulsions. Tween 20 is a non-ionic surfactant which has a molecular weight of 1228, a critical micelle concentration (CMC) of about 0.004 wt% and a HLB number of 16.7.

Sample Preparation

A surfactant solution was prepared by dissolving 2 wt% Tween 20 in double distilled water. Emulsions were prepared by weighing out 10 wt% oil and 90 wt% of the surfactant solution and sonicating for 10 - 60 seconds. Each emulsion was then diluted with either distilled water, or 2 wt% Tween 20 solution, to form a range of initial droplet concentrations (c_0 = 0 - 0.05 wt%). All the emulsions were stored in a temperature controlled environment (30 ± 1 °C, 14 days) with continuous swirling (100 rpm), and samples were withdrawn at regular intervals for analysis of droplet size and concentration using the light sattering technique described below.

Droplet Size measurements

A static light scattering technique (Horiba LA-700, Horiba Instruments Incorporated, Irving, CA) was employed to measure the droplet size distributions. This technique is similar to the one described in an earlier study (McClements and Dungan 1995). The device calculates the droplet size distribution by measuring the angular dependence of the intensity of light scattered from the droplets. The instrument used a relative refractive index of 1.10 (the ratio of the refractive index of the oil to that of the aqueous phase) to calculate the droplet size distribution from the measured light intensities.   To prevent multiple scattering effects each emulsion was diluted with distilled water prior to analysis to give a final droplet concentration of about 0.02 wt%.

Concentration measurements

The turbidity τ _of the emulsions was measured at a wavelength of 632.8 nm using the same instrument as for the droplet size determinations (Horiba LA-700). The turbidity is defined as: $\tau = \ln(I_0/I)/l$, where I_0 is the intensity of the incident light, I is the intensity of the transmitted light and l is the sample pathlength. Calibration experiments were carried out for each oil using a series of emulsions with different droplet concentrations to establish the relationship between emulsion turbidity and droplet concentration. Emulsions were prepared for each oil which had different droplet concentrations (0 - 0.05 wt%) and median droplet sizes (0.2 - 2 μm). For each size a linear relationship was found between the turbidity and the droplet concentration, the slope of the lines depending on the droplet diameter. Because the droplet size of the emulsions changed during the experiments it was necessary to develop empirical equations to calculate droplet concentration from the measured droplet diameter and turbidity: $c_0 = \tau f(d_m)$. _We found that the following empirical function gave a good agreement with our experimental data (r > 0.97):

$$f(d_m) = [A + B \ln d_m]^{-1} \tag{1}$$

Where A and B are constants which depend on the type of oil used. The values of A ranged from 37 to 47, and B from 20 to 28, when d_m was expressed in micrometers. It must be pointed out that this relationship was developed for a series of emulsions with fairly narrow size distributions. During the solubilization process the droplet size distributions of many of the emulsions broadened dramatically, and some of them eventually became bimodal, which may lead to appreciable errors in the accuracy of the above relationship. Nevertheless, it is expected that the absolute error will be relatively small because the droplet concentration is so low in the latter stages of the experiments.

RESULTS AND DISCUSSION

Affect of Droplet Concentration

The results of the light scattering study of solubilization in *n*-hexadecene emulsions with different initial droplet concentrations are shown in Figures 2 and 3. When *n*-hexadecene emulsion droplets were suspended in pure water the droplet concentration remained constant with time, as did the droplet size distribution. On the other hand, when they were suspended in a micellar solution (2 wt% Tween 20) the droplet concentration decreased with time, indicating that solubilization occurred, *i.e.*, *n*-hexadecene molecules moved from the emulsion droplets into the surfactant micelles. At sufficiently long times, the concentration of emulsion droplets reached a constant value, which indicated that either the micelles have become fully saturated, or that all of the emulsion droplets had been completely solubilized, depending on whether the initial concentration of oil in the system was above or below the amount needed to saturate the micelles.

Intuitively, one might expect the size of the emulsion droplets to shrink with time as solubilization proceeded, because oil molecules move from the droplets into the surfactant micelles. Nevertheless, our measurements clearly show that there was actually an increase in the median droplet diameter with time, the affect being most pronounced in the dilute emulsions (Figures 2 and 3). It has recently been postulated that this net increase in droplet diameter with time is because of enhanced Ostwald ripening in the presence of surfactant micelles (Weiss *et al* 1996a,b). When the smaller droplets shrink below a certain critical size

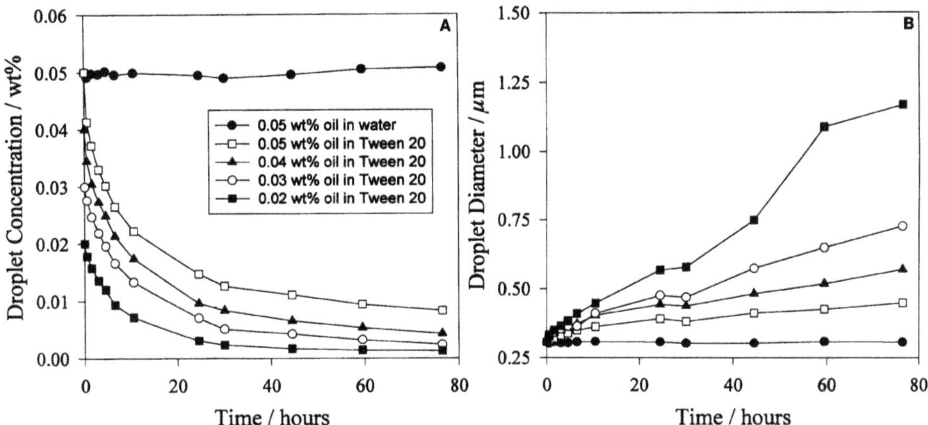

Figure 2. Dependence of droplet concentration and diameter on time for *n*-hexadecene oil-in-water emulsions suspended in either distilled water or 2 wt% Tween 20 solution (initial d_m = 0.30 μm).

their water-solubility increases dramatically and so they rapidly dissolve, redistributing their contents amongst the remaining larger droplets. The net increase in droplet size with time is probably because droplet growth due to Ostwald ripening outweighs droplet shrinkage due to solubilization. This mechanism would also account for the observation that the increase in droplet size with time occurs more quickly in the emulsions with the smaller initial droplet diameters: when a droplet dissolves in a dilute emulsion there are fewer droplets remaining to redistribute the oil amongst, and so the fractional increase in size of each of the remaining droplets is larger.

Recently, a mathematical equation has been derived to model the solubilization process of emulsion droplets dispersed in a micellar solution (McClements and Dungan 1995):

$$\frac{dc_{aq}}{dt} = k_i \frac{6\rho_E}{d_{32}\rho_0^2} (c_0 - c_{aq})(c_{sat} - c_{aq})$$

[2]

Here c_0 is the initial mass fraction of oil in the emulsion droplets, c_{aq} is the mass fraction of oil solubilized in the aqueous micellar solution at time t, c_{sat} is the maximum mass fraction of oil that can be solubilized by the micellar solution at saturation, ρ_E is the density of the emulsion, ρ_0 is the density of the oil, d_{32} is the volume-surface mean droplet diameter and k_i is the interfacial mass transfer coefficient. The interfacial mass transfer coefficient represents the mass of oil which flows out of the droplets per unit surface area per unit time, and has units of kg m^{-2} s^{-1}. This equation shows that the rate at which oil moves from an emulsion droplet into a surfactant micelle depends on the concentratation of unsaturated micelles remaining in the aqueous phase ($c_{sat} - c_{aq}$), the concentration of unsolubilized oil in the system ($c_0 - c_{aq}$) and the surface area of oil exposed to the aqueous phase.

Affect of Initial Droplet Size

The objective of this series of experiments was to study the influence of the initial droplet diameter on the kinetics of solubilization. A series of emulsions were prepared with the same

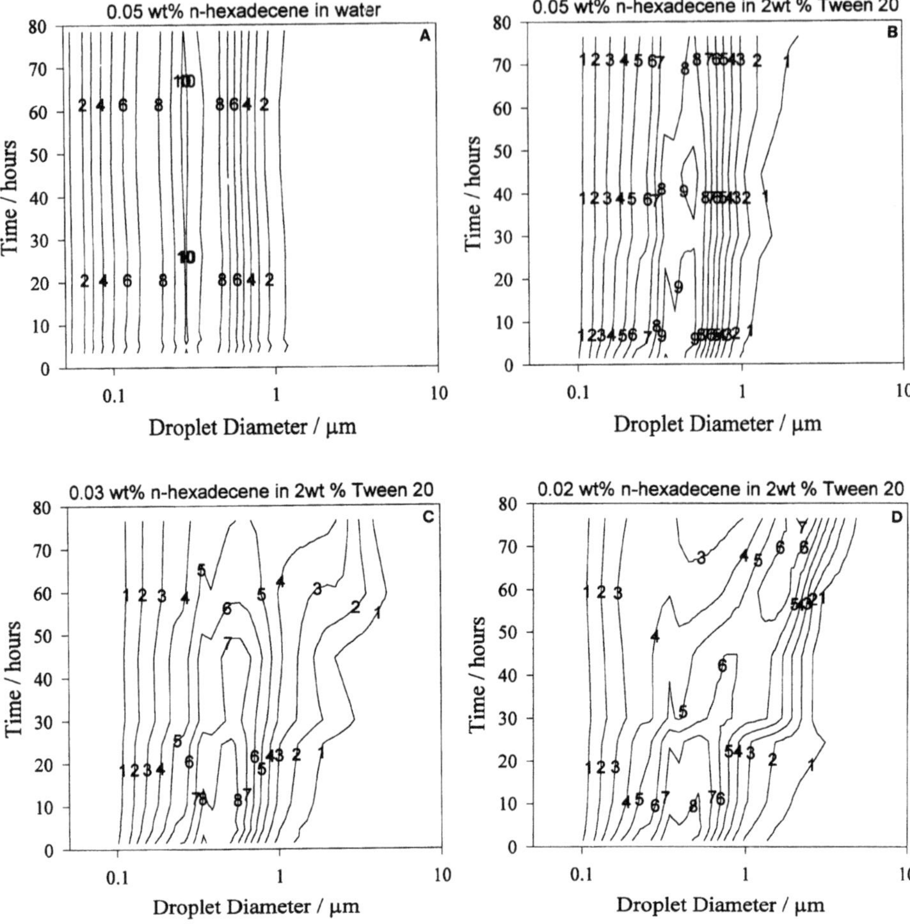

Figure 3. Variation in the droplet size distribution with time for four n-hexadecene oil-in-water emulsions with the same initial droplet diameter (d_m = 0.30 μm), but different droplet concentrations.

initial composition (0.03 wt% n-hexadecane in 2 wt% Tween 20 solution), but with different mean droplet diameters (0.17, 0.30, 0.59 and 0.73 μm). The change in droplet concentration and size distribution were then measured as a function of time (Figure 4). Solubilization was observed to occur more rapidly in the emulsions containing smaller droplets. This is because smaller droplets have a larger surface area of oil exposed to the aqueous phase. We found that a value of $k_i = 6.0 \times 10^{-7}$ kg m^{-2} s^{-1} gave a good description of all our experimental results, which is in good agreement with previous studies of hydrocarbon oils solubilized by non-ionic surfactant micellar solutions (Weiss $et\ al$ 1996a,b).

The growth of the emulsion droplets with time also depended on the initial size of the emulsion droplets (Figure 4). To represent the data from emulsions with different initial droplet diameters on the same figure the droplet diameters were normalized: $(d_t - d_0)/d_0$ where d_0 and d_t are the initial median droplet diameter and the corresponding value after time t. The data clearly shows that smaller droplets grow more rapidly than larger droplets, which suggests that the increase in droplet diameter with time due to Ostwald ripening is more pronounced for smaller droplets.

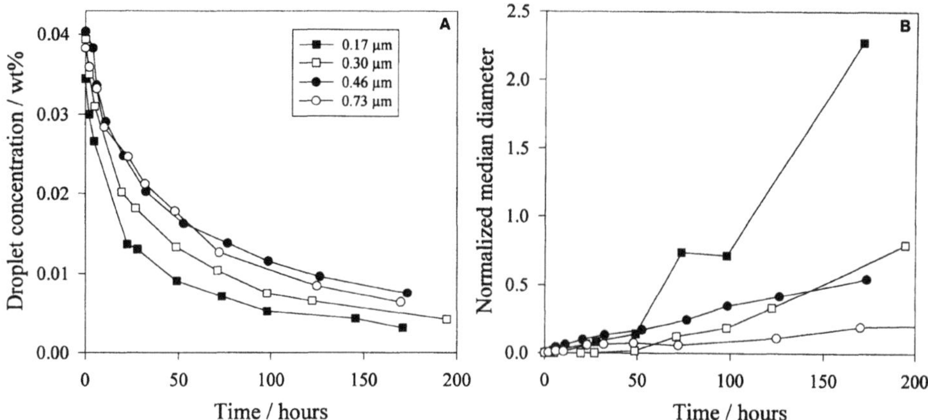

Figure 4. Comparision of the solubilization kinetics of 0.03 wt% n-hexadecane oil-in-water emulsions with different mean droplet diameters.

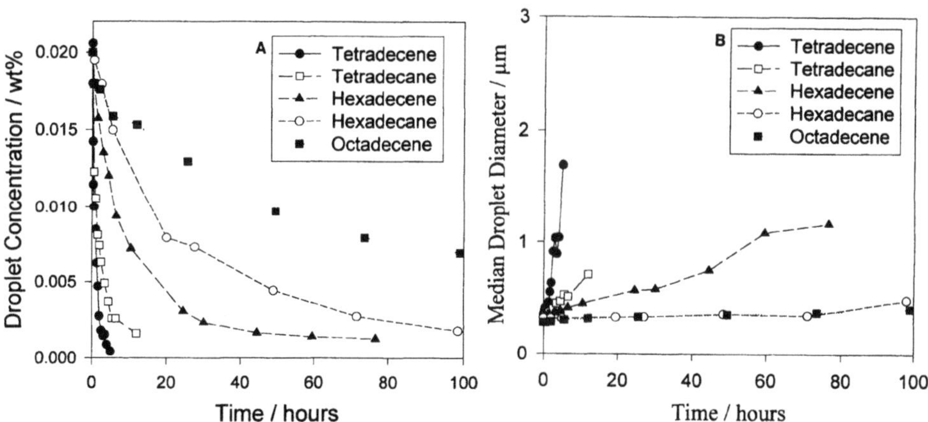

Figure 5. Comparision of the solubilization kinetics of a series of 0.02 wt% hydrocarbon oil-in-water emulsions with the same initial droplet diameter (0.3 µm), but containing different hydrocarbon oils.

Affect of Molecular Structure of Oil

The affect of the chemical structure of the hydrocarbon molecules on the rate of solubilzation in a series of 0.05 wt% oil-in-water emulsions made from different oils is illustrated in Figure 5. As the chain length of the hydrocarbons decreases the solubilization rate increases, *i.e.,* the decrease in droplet concentration with time is more rapid. In addition, the increase in droplet diameter with time occurs more quickly in emulsions with shorter chain hydrocarbons. The rate of solubilization was also observed to occur more rapidly for unsaturated than for saturated oils.

To attempt to understand the relationship between the interfacial mass transport rate and the molecular structure of the oil molecules in a droplet we consider the results of a recent molecular dynamics simulation (Esselink *et al* 1994). This simulation suggests that there are at least three possible mechanisms by which oil molecules can be transferred from an emulsion droplet into a surfactant micelle:

(i)   direct solubilization of oil molecules in water, then subsequent uptake by micelles in the aqueous phase;

(ii)   collision of a micelle with the surface of an emulsion droplet, leading to incorporation of oil molecules into the micelle;

(iii)   spontaneous "budding-off" of a number of oil and surfactant molecules from the surface of a droplet to form a micelle.

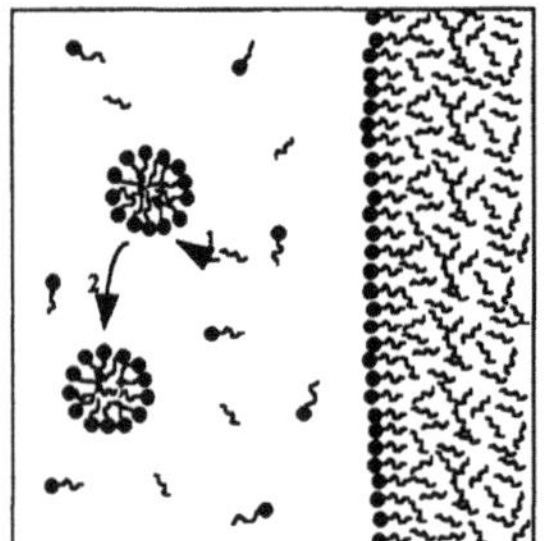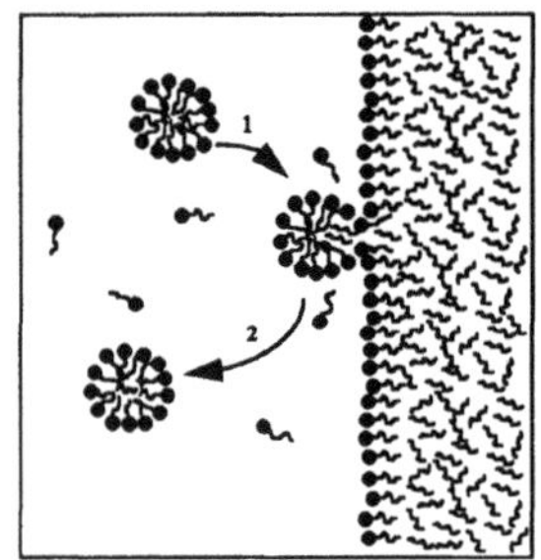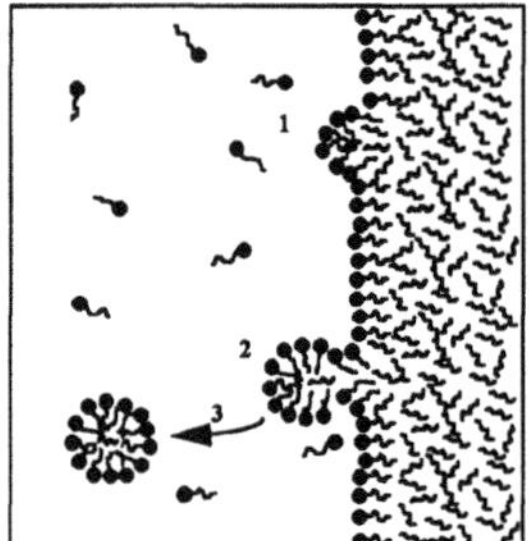

Figure 6. Three possible mechanisms for oil solubilization by surfactant micelles.

We will consider the relative importance of each of these mechanisms in turn. The water-solubility and molecular diffusion of hydrocarbons increases as their chain length decreases or they become more unsaturated (Israelachvili 1992), which would lead to an increase in the rate of solubilization as observed in our experiments. The water-solubility of medium chain hydrocarbons is so low that one would not expect mechanism (i) to cause significant Ostwald ripening under normal conditions. Nevertheless, this mechanism may become important once emulsion droplets have shrunk below their critical size and dissolve rapidly. In mechanisms (ii) and (iii) oil molecules are "pulled" from the surface of an emulsion droplet when a micelle collides with the droplet, or when a micelle "buds off". The energy required to pull a hydrocarbon molecule from the interior of a droplet into a micelle will be expected to increase as its size increases which again may account for the experimental observation that solubilization occurs more quickly in droplets containing shorter chain hydrocarbons.

The ability of a surfactant micelle to incorporate non-polar molecules also depends on the size of the hydrophobic interior of a micelle relative to that of the hydrocarbon molecules. If the length of a hydrocarbon molecule is sufficiently greater than the length of the surfactant tail then a micelle may not be able to accommodate the hydrocarbon without significantly changing its shape, which may be thermodynamically unfavourable. Thus a given micellar system may only be able to solubilize molecules below a certain size. This has important implications for the development of micellar systems for selective extraction of specific molecules from a complex mixture of non-polar molecules. For example, it may be possible to selectively extract molecules below a certain size from a droplet which contains a mixture of molecules.

Solubilization of a Food Oil

So far all the work that has been discussed has used simple hydrocarbons as the oil phase of the emulsions. Most edible oils consist of a complex mixture of triacylglyercols, and so we have carried out a series of experiments using a real food oil. Measurements of the change in droplet size and concentration in corn oil-in-water emulsions containing droplets suspended in an aqueous micellar solution indicated that there was no appreciable solubilization over a 200 hour period (Figure 7). This could have been for two reasons, either the kinetics of the solubilization process is very slow, or the maximum amount of oil which can be solubilized in the micelles is very low.

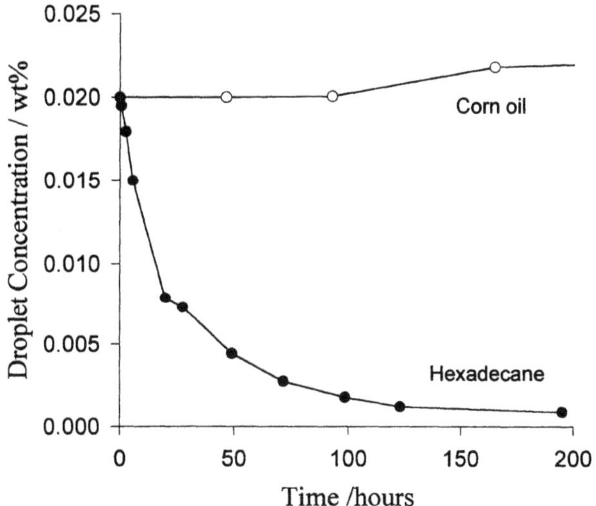

Figure 7. Comparision of the solubilization kinetics of a 0.05 wt% oil-in-water emulsions containing either hexadecane or corn oil (initial droplet diameter 0.3 μμ)

Due to the large size of the triacylglycerol molecules compared to the hydrophobic core of the surfactant micelles we believe that the micelles used in this study were not large enough to solubilize significant amounts of triacylglycerol molecules. Nevertheless, it should be possible to develop micellar systems which can be used to solubilize triacylglycerol molecules, by varying the environmental conditions (pH or temperature), adding co-surfactants (such as short chain alcohols) or using surfactants with different molecular structures. Our laboratory is currently working on this area, and hopes to develop novel methods of processing foods using micelles based on our findings.

CONCLUSIONS

The ability of surfactant micelles to solubilize and transport non-polar molecules has important applications in the food industry, including fractionation of oils, controlled ingredient release and controlled chemical reactions.   To fulfill the potential of micellar technology in the food industry it is important to understand the factors which affect the rate of oil solubilization in surfactant micelles.   In this chapter, I have reviewed some of the work carried out in my laboratory about the effects of the molecular structure of emulsified oils on the rate and extent of their solubilization by non-ionic surfactant micelles.   The transfer of oil molecules from emulsion droplets to surfactant micelles depends on their molecular structure: the rate of solubilizaton increasing as the chain length of the hydrocarbons decreases, and as the hydrocarbons become more unsaturated.   This may have important consequences for the practical application of micelles in various processing operations.

REFERENCES

Aveyard, R., Binks, B.P., Clark, S., and Fletcher, P.D.I., 1990, Cloud points, solubilization and interfacial tensions in systems containing nonionic surfactants. *Chem. Tech.Biotech*, 48:161.

Carroll, B.J., 1981, The kinetics of solubilization of nonpolar oils by nonionic surfactant solutions, *J. Colloid Interface Sci.*, 79:126.

Coupland, J.N. Weiss, J., Lovy, A., and McClements, D.J., 1996, Comparison of the solubilization kinetics of triacyl glycerol and hydrocarbon emulsion droplets in a micellar solution. *J. Food Sci.* Submitted.

Dickinson, E., 1992, *An Introduction to Food Colloids*, Oxford Science Publications, Oxford.

Dickinson, E., and McClements, D.J., 1995, *Advances in Food Colloids*, Blackie Academic and Professional, London.

El-Nokaly, M., Hiler, G., and McGrady, J., 1991, Solubilization of water and water-soluble compounds in triglycerides, in: *Microemulsions and Emulsions in Foods*, El-Nokaly, M., and Cornell, D., eds., Americal Chemical Society, Washington DC.

Elworthy, P.H., Florence, A.T., MacFarlane, C.B., 1968, *Solubilization by surface-active agents and its Applications in chemistry and biological Sciences*, Chapman and Hall, London.

Esselink, K,. Hilbers, P.A.J., van Os, N.M., Smit, B. and Karaborni, S., 1994, Molecular dynamics simulations of model oil/water/surfactant systems, *Colloids and Surfaces A*, 91:155.

Han, D., Yi, O.S., and Shin, H.K., 1990, Antioxidative effect of ascorbic acid solubilized in oils via reversed micelles, *J. Food Sci.*, 55:247.

Hiemenz, P.C., 1986, *Principles of colloid and surface chemistry*, Marcel Dekker, New York.

Israelachvili, J.N., 1992, *Intermolecular and Surface Forces*, Academic Press, London.

Krog, N.J., 1990, Food emulsifiers and their chemical and physical properties, in: *Food Emulsions*, Larsson K., and Friberg, S.E. eds., Marcel Dekker, New York.

Mackay, R.A., 1987, Solubilization, in: *Nonionic Surfactants: Physical Chemistry*, Schick, M.J., ed., Marcel Dekker, New York.

McBain, M.E.L., 1995, *Solubilization and Related Phenomena*, Academic Press, New York.

McClements, D.J., Dungan, S.R., 1993, Factors that affect the rate of oil exchange between oil-in-water emulsions droplets stabilized by a nonionic surfactant: Droplet size, surfactant concentration and ionic strength. *J. Phys. Chem.* 97:7304.

McClements, D.J., and Dungan, S.R., 1996, Light scattering study of solubilization of oil-in-water emulsion droplets in a nonionic surfactant solution. *J. Collloid Interface Sci.*. In press.

McClements, D.J. , Dungan, S.R. , German, J.B., and Kinsella, J.E., 1992, Oil exchange between oil-in-water emulsion droplets stabilized with a nonionic surfactant. *Food Hydrocolloids*, 6:415.

McNulty, P.B. 1987, Flavour release - elusive and dynamic, in: *Food Structure and Behaviour* Blanshard, J.M.V., and Lillford, P., Academic Press, London.

McNulty, P.B. and Karel, M., 1987, Factors affecting flavor release and uptake in O/W emulsion, *J. Food. Tech.* 8: 309, 8:319, 8:415.

Meguro, K., Ueno, M., and Esumi, K., 1987, Micelle formation in aqueous media, in: *Nonionic Surfactants: Physical Chemisty*, M.J. Schick, ed., Marcel-Dekker, New York.

Nakagawa, T., 1966, Solubilization, in: *Nonionic Surfactants,* Schick, M.J., ed., Marcel Dekker, New York.

Ward, A.J.I., 1989, The kinetics of oil solubilization in aqueous micellar solutions, *Proc. Irish Acad.* 89B:375.

Ward, A.J.I., and Quigley, K., 1990, Solubilization of nonpolar oils in aqueous micellar mixtures of ionic and nonionic surfactants, *J. Disp. Sci. Tech,* 11:143.

Wedzicha, B.L., 1988, Distribution of low-molecular-weight food additives in dispersed systems, in: *Advances in Food Emulsions and Foams,* Dickinson, E. and Stainsby, G., eds., Elsevier Applied Science, London.

Weiss, J., Coupland, J.N. and McClements, D.J., 1996a, Solubilization of hydrocarbon emulsion droplets suspended in nonionic surfactant micelle solutions. *J. Physical Chemistry.* In press.

Weiss, J., Coupland, J.N. and McClements, D.J., 1996b, Molecular structure of hydrocarbons in emulsion droplets affects their solubilization in nonionic surfactant micelles. *J. Colloid Interface Sci.* Submitted.

Wolcott, T., 1995, New separation and ingredient-release technology, *Prepared Foods,* 71:164.

PROPERTIES OF LOW-FAT, LOW-CHOLESTEROL EGG YOLK
PREPARED BY SUPERCRITICAL CO_2 EXTRACTION

Neal A. Bringe

Ceregen, A Unit of Monsanto Company,
700 Chesterfield Parkway North, St. Louis, MO 63198

SUMMARY

A dry egg yolk ingredient called Eggcellent[TM] has 74% less fat and 90% less cholesterol than liquid egg yolks, when reconstituted on an equal protein basis. The phospholipids and proteins are retained, enabling the ingredient to have the taste and texturizing properties of fresh egg yolk. Using the new yolk, it is possible to significantly improve the acceptability of low-fat, low-cholesterol bakery products, scrambled eggs and mayonnaise dressings without losing nutritional claims. The structures and functional properties of egg yolk components and the conditions required to optimize their benefits in foods are reviewed. The lipoproteins of low-fat, low-cholesterol yolk have valuable properties as flavorants, texturizers, foaming agents, emulsifiers, antioxidants, colorants, and nutraceuticals.

INTRODUCTION

A new egg yolk ingredient was developed at The NutraSweet Company, Box 730, 1751 Lake Cook Rd., Deerfield, IL 60015 to meet the needs for a low-fat, low-cholesterol egg yolk (LFLC yolk). This ingredient, called Eggcellent[TM] or LFLC yolk, is produced by using supercritical CO_2 to extract triacylglycerols and cholesterol, retaining the valued proteins and phospholipids which made up 59% and 24% of the ingredient (dry basis), respectively (Bringe & Cheng, 1995). The special properties and benefits of egg yolk proteins and phospholipids, which are complexed as lipoproteins, depends on their unique structures.

STRUCTURE OF EGG YOLK LIPOPROTEINS

The main lipoprotein structures of egg yolk are shown in Fig. 1. The small low density lipoprotein (LDL) particles contain 90% of the yolk lipid (Woodward, 1990). These particles have a triacylglycerol core (61.5%) surrounded by phospholipids (23%), cholesterol (3.5%) and 6 different vitellenin apoproteins (12%)(Burley and Vedhra, 1989). The vitellenins are 23% of yolk protein (Woodward, 1990). Vitellenin I (also called apo-VLDL-II), is about 32% of the LDL proteins and seems to be synthesized specifically to stabilize egg yolk LDL-particles by associating with phospholipids (Burley and Vedhra, 1989; Griffin, 1992).

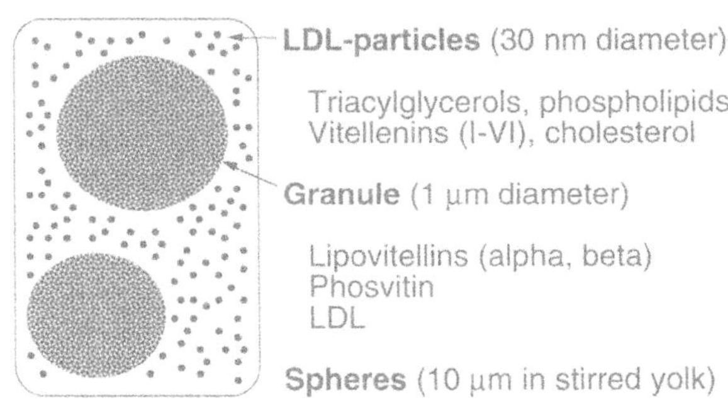

Figure 1. Native egg yolk structure.

Egg yolk granules contain half of the yolk protein, 70% of yolk calcium and 95% of yolk iron (Burley and Vedhra, 1989). The granules are 70% lipovitellin microparticles, 16% phosvitin and 12% LDL particles. The microparticles consist of vitellins (77%), phospholipids (13.3%), triacylglycerols (8.2%) and other lipids including cholesterol (1%) (Banaszak, 1991). The microparticles are associated with strands of the highly phosphorylated phosvitin proteins, and together these components surround LDL particles within the granules (Chang et al., 1977). The granules are held together by non-covalent bonds so changes in salt concentration or pH from that of native yolk, causes the granules to swell and dissociate (Burley and Cook, 1961; Chang et al., 1977; Causeret et al., 1991, 1992).

Some of the LDL-particles, granules and plasma are also present within structures called spheres, which are remnants of adjoining polyhedral grains (4-150 µm) in uncracked yolk. The median diameter of the spheres which survive mixing is about 10 µm (Fig. 2). The spheres in stirred yolk are visible in micrographs of heat-induced yolk gels (Woodward and Cotterill, 1987).

The main soluble proteins of egg yolk are the livetins (30% of yolk protein) which correspond to serum proteins. There are also small amounts of enzymes (e.g. cholinesterase, pyruvate kinase), vitamin-binding proteins (e.g., riboflavin- and retinol-binding proteins) and albumen proteins in native yolk (Burley and Vadehra, 1989). Commercial egg yolk is diluted with egg white, causing yolk solids to decrease from 51.8% to 44%.

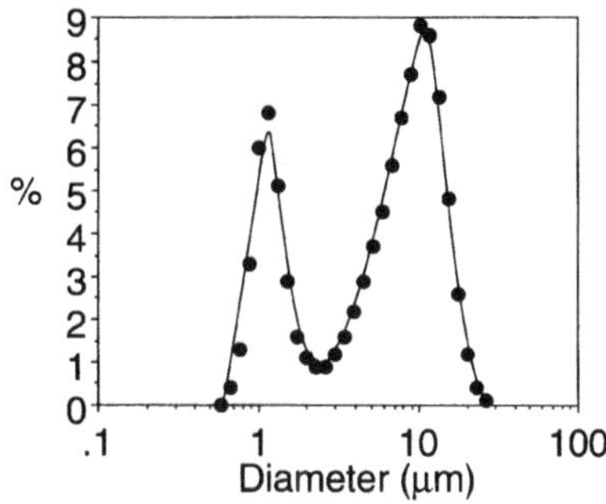

Figure 2. Particle size distribution of native egg yolk. The figure shows the volume frequency of particles at each diameter as determined by laser light diffraction using an LA-900 instrument (Horiba Instruments, Irvine, Calif.). Dilute egg yolk was sonicated (40 Watts, 150 mL) in the instrument mixing chamber for 2 min before circulating through the cuvette. The refractive index relative to water, used in calculations was 1.07.

Table 1. Egg Yolk Phospholipids and Sphingolipid. Data from Cook and Martin (1969).

	(%)
Phosphatidylcholine	77
Lysophosphatidylcholine	2.5
Sphingomyelin	2.3
$-CH_2-CH_2-N^+(CH_3)_3$	
Phosphatidylethanolamine	18
$-CH_2-CH_2-NH_2$	

The phospholipids in egg yolk are primarily phosphatidylcholine and phosphatidylethanolamine (PE)(Table 1). Phosphatidylcholine, lysophosphatidylcholine and sphingomelin have a choline group, a valued nutrient. PE has a free amino group which can participate in non-enzymic browning reactions.

When dried egg yolk is extracted, using supercritical carbon dioxide, there are no changes in the size and morphology of the granules as determined from electron micrographs and laser light diffraction measurements of particle size (Bringe and Cheng, 1995). The LDL-particles are altered because they contain most of the yolk neutral lipids which are extracted. The phospholipids and proteins of the extracted LDL-particles, form new membrane structures: rod shaped (Fig. 3, left) and vesicles (Fig. 3, right).

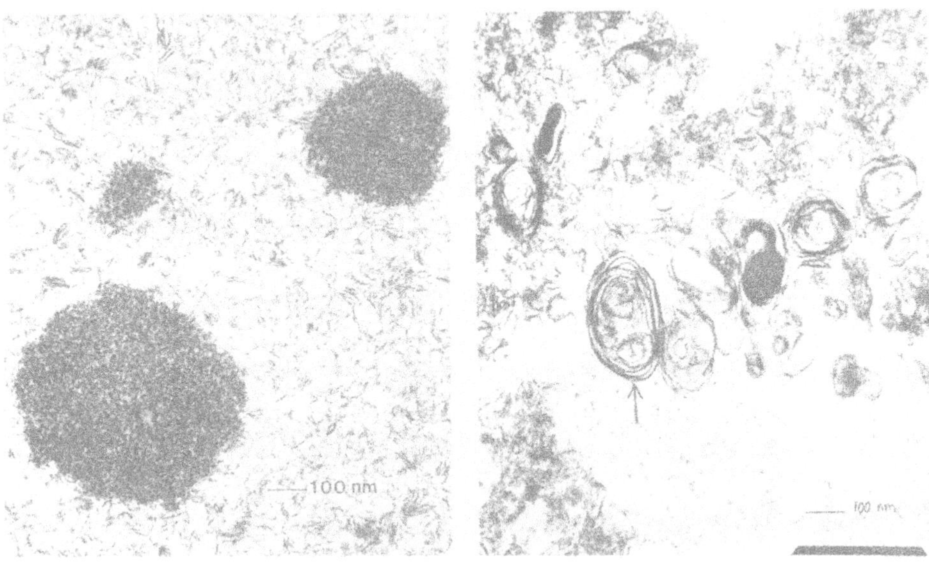

Figure 3. Transmission electron micrographs of LFLC yolk showing short rods (left) and vesicles (right), believed to be protein and phospholipid remnants of LDL-particles. Small granules are also shown (left). Samples were prepared as described before (Bringe and Cheng, 1995).

PROPERTIES OF NATIVE AND EXTRACTED EGG YOLK

Knowledge about the properties of native and extracted egg yolk, which largely reflect the properties of yolk lipoproteins, can help us to use these ingredients effectively in food products.

Rehydrate, hold water

LFLC yolk hydrates well under conditions which physically break up the powder particles. Fig. 4 shows the viscosities of aqueous suspensions of extracted yolk powder which were mixed for different lengths of time. The sharp increase in viscosity after 5

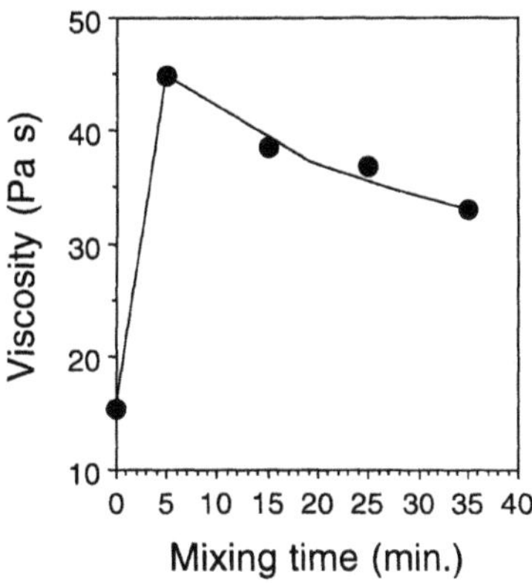

Figure 4. Viscosity of LFLC yolk as related to time of mixing. LFLC yolk was mixed using a Stephan cooker (34% solids, 5°C, under vacuum, 1,800 rpm) and the viscosity of the samples were determined at 5°C, at a shear rate of 14.6 sec^{-1} and 5 min of shear (C14 cup and bob system, VOR rheometer, Bohlin Instruments, Inc., Cranbury, N.J.).

minutes of mixing was caused by hydration of the powder. The subsequent decreases in viscosity were caused by reduction of hydrated aggregates in the reconstituted yolk.

The viscosity of reconstituted LFLC yolk is more viscous at practical concentrations (0-16% protein) than native yolk (Bringe and Cheng, 1995) because the extracted LDL structures hold more water than native LDL structures, which primarily hold fat. This means that additional thickeners are not needed when extracted yolk is substituted for native yolk on a equal or reduced protein basis in applications such as low-fat ice cream.

React with oxygen and glucose

Egg yolk lipoproteins react to a limited extent with oxygen and reducing sugars such as glucose of eggs. These oxidation and non-enzymic browning reactions are both a limitation to control and a benefit of yolk lipoproteins.

Liquid egg yolk is stable to lipid oxidation (Pike and Peng, 1985) even though it contains iron, copper, and manganese (60, 1.5 and 1 ppm, respectively) and polyunsaturated fatty acids (14-20% of the fatty acids) (Cotterill and Glauert, 1979). This protection is provided by yolk proteins (Pike and Peng, 1988a) which limit exposure of the lipids to oxygen (Burley and Vadehra, 1989), and chelate iron catalysts (Lu and Baker, 1986; Yamamoto et al., 1990).

The protective effects of yolk proteins are reduced when lipoprotein structures are disrupted, for example by protein hydrolysis (Pike and Peng, 1985), acidification (Pike and Peng, 1988b), cooking (Verstrate, 1980), high amounts of sugar added before drying (Kline et al., 1964) and by extracting neutral fats (Warren et al., 1991).

The rate of lipid oxidation in LFLC yolk was determined by measuring the changes in the amount of thiobarbituric acid reactive substances (TBARS) in the powder over several months. The rate of oxidation was negative in frozen powder, because the measured oxidation products were lost more rapidly than they were formed during frozen storage (Fig. 5). The rate of oxidation increased exponentially with increases in temperature, so the positive rates can also be shown as a straight line in a semi-log plot (Fig. 6). Also shown in Fig. 6 is the effect of powder density. Porous flakes are created by extracting vacuum-drum-dried egg yolk. When these flakes are reduced in size, the product oxidizes at much slower rates because dense powders limit exposure of lipids to oxygen. Lipid oxidation, therefore, can be controlled by freezing or by increasing powder density.

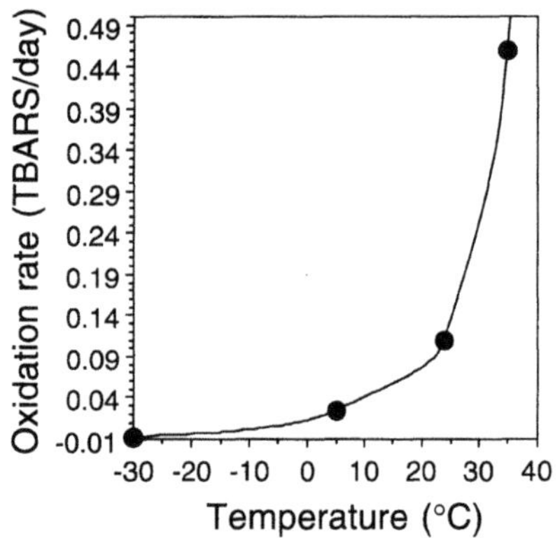

Figure 5. LFLC yolk oxidation rate as related to temperature. TBARS were determined by the method of Tarladgis et al. (1960). The relative vapor pressure of the powder was 0.2 (23°C). Powder was stored in heat-sealed high barrier packages (3.75 mil 100 ga. nylon/pvdc/0.0025 modified polyethylene) containing air.

The TBA reactive substances, which are aldehydes such as malondialdehyde, are the precursors to flavor compounds people perceive as oxidized or cardboard flavor. This hypothesis is supported by the observation that the TBARS of extracted yolk decreased during frozen storage as low levels of oxidized flavor attribute appeared in scrambled eggs made with the powder (Fig. 7). The eggs at the end of this storage test, had good overall flavor, and continue to maintain that flavor with longer storage because the off-flavor precursors have been used up.

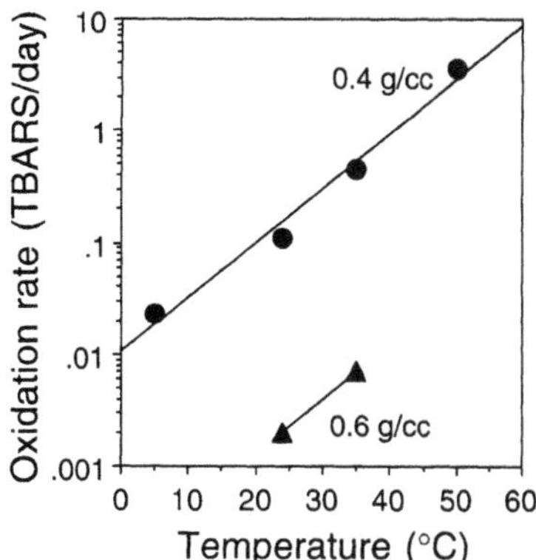

Figure 6. LFLC yolk oxidation rate as related to packed-bulk-density and temperature. $R^2 = 0.982$ for linear regression of data for powder with 0.4 g/cc.

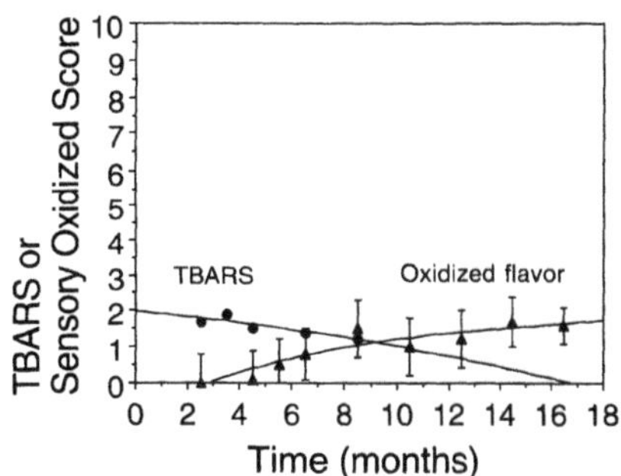

Figure 7. Oxidized flavor of scrambled eggs and TBARS of LFLC yolk as related to storage time of LFLC yolk. Scrambled eggs were made the day before sensory panels using LFLC yolk stored various periods of time at -29°C in nitrogen flushed, heat-sealed high barrier packages (3.75 mil 100 ga. nylon/pvdc/0.0025 modified polyethylene). Five to nine trained consumers from Mt. Prospect, IL area evaluated the cooked eggs using a descriptive flavor and texture profile method using a 15 point intensity scale. Means were adjusted (least squares) for any imbalance in the design. The increase in cardboard/oxidized flavor over the 16.5 months of storage showed a statistically significant linear trend ($p < 0.10$). Fresh eggs had a sensory score for oxidized flavor of 0.6 ± 0.2. TBARS values after 8.5 months were reported as <1.0.

Frozen storage can also limit non-enzymic browning reactions in egg yolk powders. The effects of temperature on the rates of non-enzymic browning in full fat dried egg yolk was determined by comparing the changes in the Hunter Colorimeter "a" value compared to frozen controls. The redness or brown color increased much more rapidly at 23°C than at 5°C (Fig. 8). No change in color occurred in the frozen product.

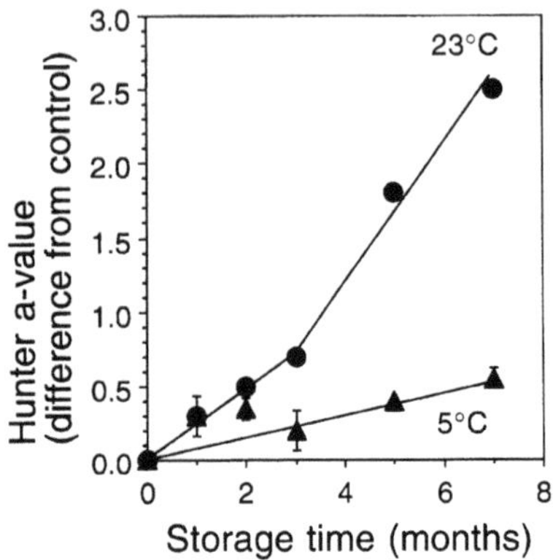

Figure 8. Non-enzymic browning of dried egg yolk as related to storage time and temperature.

The rate of browning in dried egg yolk was partially facilitated at warmer temperatures by a collapse of the powder structure. The collapse of the powder structure was indicated by increases in the relative vapor pressure of the dried yolk, compared to the frozen control (Fig. 9). Similar observations were made for LFLC yolk. Protein-protein interactions replace protein-water interactions creating more free water. The relative vapor pressure of the powders increased most at 23°C because this temperature is more above the glass transition temperature of the yolk polymers allowing greater structural changes. Collapse of the powder structure facilitates browning reactions by bringing reactants closer to each other (Buera and Karel, 1995). The resulting increases in relative vapor pressure are undesirable because high relative vapor pressure also facilitates reactions such as lipid oxidation. Consequently, frozen storage is recommended for storing egg yolk powders. Full-fat powders harden as one unit when frozen, so they must be tempered before use. Frozen LFLC-yolk is free flowing.

Non-enzymic browning reactions occur in dried eggs between carbonyl groups of glucose, glucose reaction products (e.g. hydroxymethyl furfural) or lipid oxidation products (e.g. malondialdehyde) and the amino groups of proteins or of PE. Brown color is not the main defect of the browning reaction. The glucose or glucose-related reaction with PE also creates an off-flavor, described as vial and nauseating (Feeney, 1989), and is a primary cause of off-flavor development in dried eggs. This finding was illustrated by a

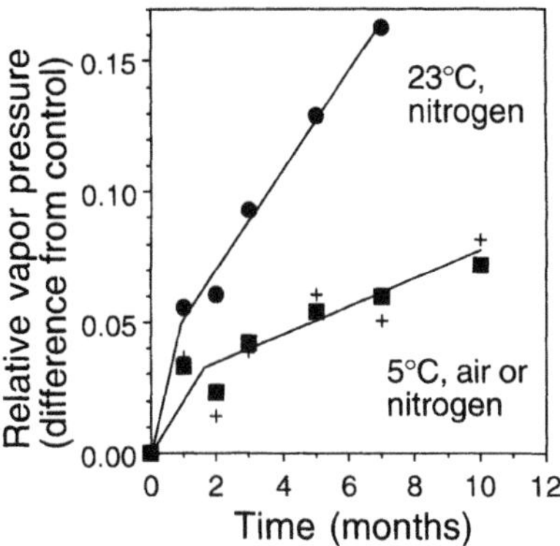

Figure 9. Relative vapor pressure of dried egg yolk as related to storage time and temperature. Vapor pressure determinations were made after powders were equilibrated at 23°C. The high barrier, heat-sealed packages kept powder moisture at 2.6% in LFLC yolk stored at 5°C or in the range 2.6-3.3% at 23°C as determined by a microwave moisture test. The packages contained either nitrogen or air as indicated in the figure. The initial relative vapor pressure of LFLC yolk was 0.30.

good correlation between the palatability of dried eggs and the amount of glucose-PE reaction products, measured by lipid flourescence (Boggs et al., 1946). Removal of glucose from liquid eggs using glucose oxidase, results in a dried product which is much more stable to off-flavor development (Kline et al., 1951a, 1951b). This off-flavor issue is also significant for dried egg yolk, but is not as great, because the yolk powders contain less glucose than whole eggs (0.4% vs. 1.2%, dry basis, respectively; Kilara and Shahani, 1973; Powrie, 1977). However, if aldehydes are allowed to form from lipid oxidation they may also form off-flavors by reacting with PE during storage.

When oxidation and non-enzymic browning reactions involving yolk lipoproteins, occur during cooking, they create valuable flavors which characterize foods such as mayonnaise, scrambled eggs, French ice cream, pound cake, custard and egg nog. Aldehydes formed from these reactions (e.g. hexanal) were positively correlated to scrambled fresh egg flavor, overall impression and sweet sensory characteristics of scrambled eggs (Warren et al., 1995). It is difficult to replicate these flavors without egg yolk lipoproteins as illustrated by comparing published consumer acceptance scores for full-fat yellow cake and scrambled eggs and low-fat products made without egg yolk (Fig. 10). The low fat products without egg yolk were optimized commercial products. A main reason they scored low compared to control products is that egg yolk lipoproteins are needed to create the desired flavors. It is possible to eliminate this gap in acceptability without losing low-fat low-cholesterol claims, by using LFLC yolk in low fat cakes and eggs. This observation was reported for LFLC scrambled eggs and mayonnaise dressing

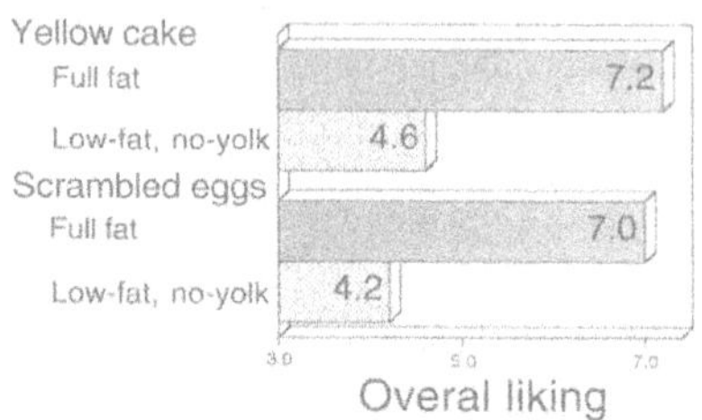

Figure 10. Acceptability of low-fat, no-yolk cake and eggs compared to full-fat, yolk containing products. Sensory tests were conducted using a 9 point hedonic scale. Data from Gardner et al. (1982) and Hahn (1993).

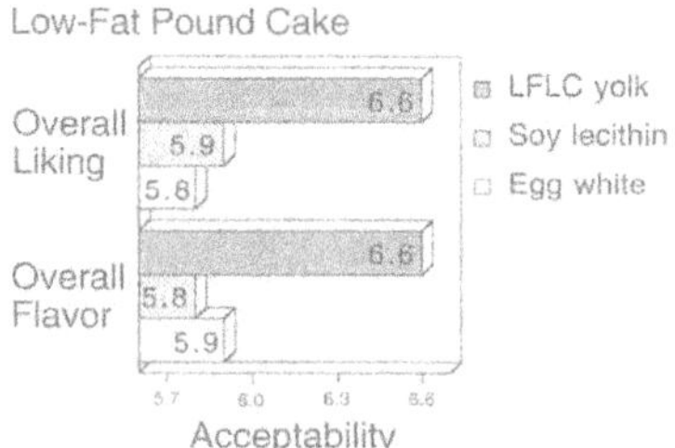

Figure 11. Acceptability of low-fat pound cake made with LFLC yolk (1 %), soy lecithin (0.23%) or additional egg white (1.0%). The dry ingredients replaced other cake solids; moisture of the batters was constant (35.5%). One-inch cake slices were presented successively in balanced order to panelists (60). Acceptability of the cake slices were rated using an hedonic scale where 1 is "dislike extremely," 5 is "neither like nor dislike" and 9 is "like extremely." The cake containing LFLC yolk was preferred over the cakes containing added egg white or lecithin (p < 0.05).

(Bringe and Cheng, 1995) and is illustrated here for low-fat pound cake. A low fat pound cake (water, sugar, flour, microparticulated whey protein concentrate, shortening, egg white [0.7%], modified corn starch, baking powder, salt, emulsifiers, hydroxypropyl methylcellulose, and xanthan gum) was made with small amounts of added LFLC yolk, egg white or soy lecithin. Highly acceptable cakes were created by added LFLC yolk, but not by adding soy lecithin or egg white (Fig. 11).

Disrupt food matrices

Egg yolk lipoproteins disrupt food matrices making food products smoother or more tender. The effects of storage time and the type of additional ingredient on the firmness of low fat pound cake slices was determined by measuring penetration force. Additional egg white caused the cakes to be more firm than control cakes because the denatured globular proteins participated in matrix formation with gluten and starch (Fig. 12). The addition of extracted yolk and soy lecithin made the cakes more tender than control cakes because lipoproteins and phospholipids disrupted the gluten matrix. Based on acceptance panel results, using 60 panelists, the texture of the cakes containing LFLC yolk were preferred ($p < 0.05$) over the texture of the cakes containing egg white or soy lecithin (6.4 v. 6.0 and 5.9, respectively, on a 9 point scale). Yolk lipoproteins contributed textural value which was difficult to replace with other proteins or phospholipids.

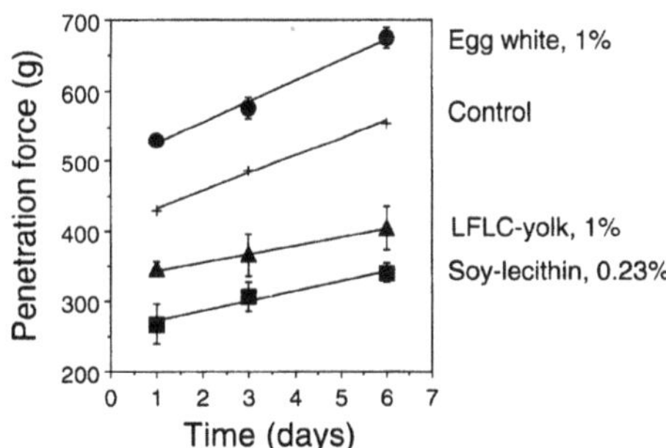

Figure 12. Firmness of low-fat pound cake as related to ingredient-type and storage time. Soy lecithin added (0.23% w/w) was equal to the amount of LFLC yolk added (1% w/w) on a phospholipid basis. Added ingredients replaced other cake solids; cake batters were 35.5% moisture. Slices from cakes were stored in sealed plastic bags. Firmness was measured using a TA.XT2 Texture Analyzer (Texture Technologies Corp., Scarsdale, N.Y.), where a flat probe (25 mm diameter) penetrated (6.3 mm, 1 mm/sec) into a slice of cake (2.5 mm thick). Values show average and standard deviation for duplicate cakes.

Stabilize air cells and fat droplets

Egg yolk lipoproteins stabilize air cells in food batters and fat droplets in sauces and dressings. The factors affecting the foaming properties of native and extracted egg yolk were determined by whipping the yolk ingredients in water using a kitchen mixer (Sunbeam Mixmaster). Native egg yolk and LFLC yolk behaved similarly, both showing a maximum foam formation at 2.5% protein and increased foam stability with increasing concentration (Fig. 13). The maximum foamability possibly relates to an optimum packing of yolk granules between air cells.

Foaming properties are desirable in food batters but it is desirable to limit foam formation when reconstituting extracted yolk. Limited foaming can be accomplished by using a vacuum processor or by reconstituting at 20% protein, 35% solids.

Optimum foaming capacity and stability occur in the pH range 6.0-6.5 (Fig. 14), which is the pH range where egg yolk granules are intact and where LDL particles have little net charge (Nakamura et al., 1982). Particles such as yolk granules help to stabilize air cells by acting as spacers between the air cells (Wasan, 1992). The foam stabilizing properties of egg yolk are reduced when granules are dissociated using NaCl (Awazuhara and Nakamura, 1986; Dyer-Hurdon and Nnanna, 1993). Deoiled soy lecithin did not create foams under these conditions (pH 4.5-7.5), indicating that the yolk foam stabilizers were lipoproteins, not phospholipids. Lecithin is a foaming agent in food batters however, because of favorable interactions with wheat proteins (Le Meste and Davidou, 1995).

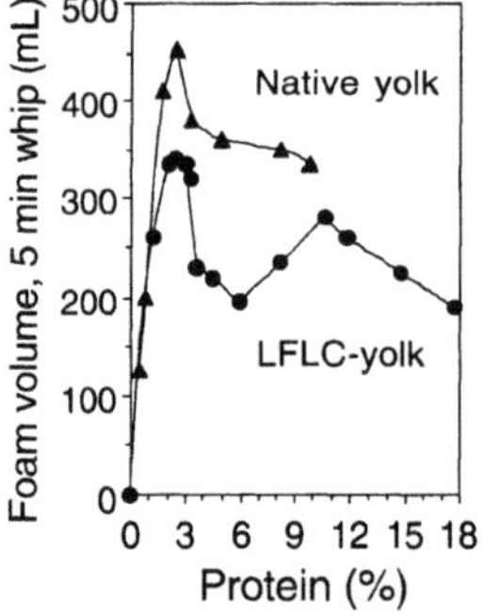

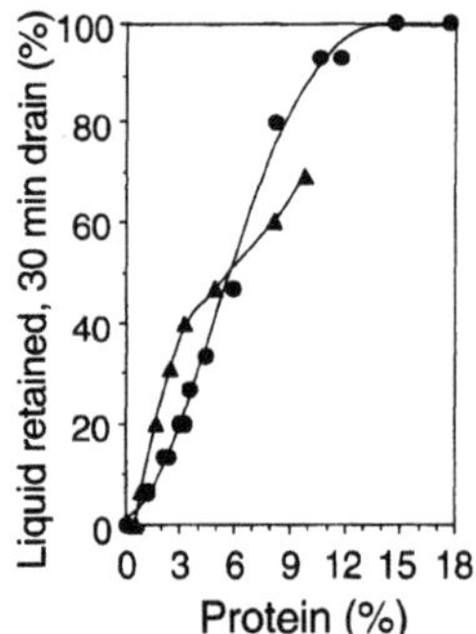

Figure 13. Foaming capacity (left) and foaming stability (right) of native and LFLC egg yolk as related to yolk protein concentration. LFLC yolk was reconsituted using a Stephan mixer (1800 rpm, 35 min). Reconstituted LFLC yolk and native egg yolk were mixed under vacuum at the desired solids levels (1500 rpm, 15 min) and adjusted to pH 7.0 using 1 N NaOH. After aging overnight (5°C), the samples were equilibrated at 22°C. Dilute yolks (75 mL) were mixed for 5 min in a Sunbeam Mixmaster (235 W, 1.43 L bowl rotated at 78 rpm) in duplicate. The foam was carefully transferred to a 500 mL graduated cylinder to measured foam volume. The percentage of liquid retained in the foams after 30 min was used as a measure of foam stability.

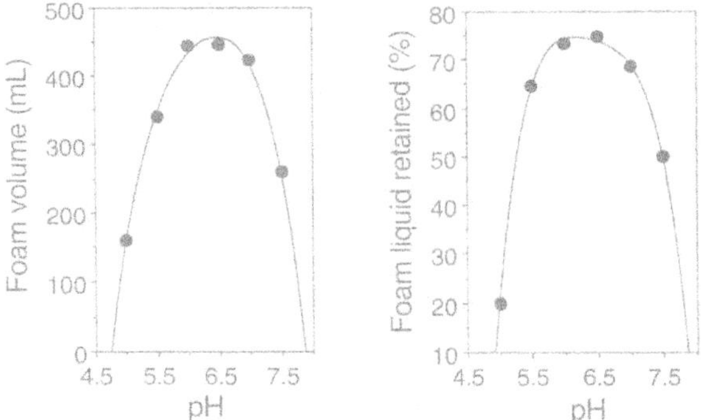

Figure 14. Foaming capacity and foam stability of dilute native egg yolk as related to pH. Yolk samples were prepared and tested as discribed in figure caption above; samples were acidified using 1N HCL. Protein concentration was 2.7%.

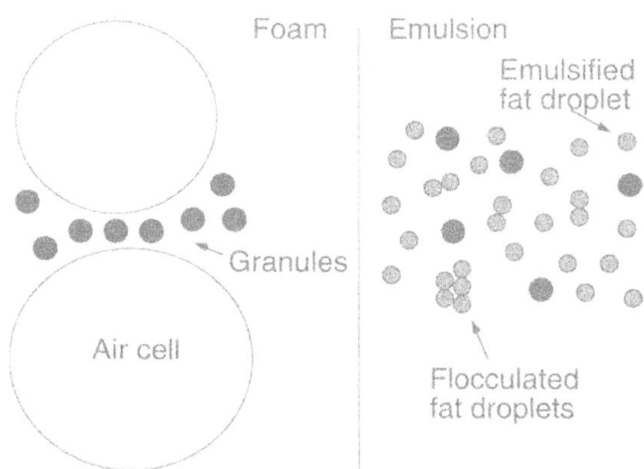

Figure 15. Comparison between the gross structures of foams and emulsions and the roles of yolk granules in both systems.

The volume of low-fat pound cake (975 ± 21 cc) measured by rapeseed displacement, was improved by adding 0.23% soy lecithin (1060 ± 14) or 1.0% LFLC yolk (1098 ± 32), equivalent additions on a polar lipid basis.

Egg yolk granules have different roles in emulsions and foams. The granules stabilize air cells which are much larger than granules, but do not stabilize fat droplets, which are smaller or the same size as granules (Fig. 15). Fat droplets are best stabilized by a cohesive protein film at the droplet surfaces. When the granules are intact, half of the yolk proteins, the granule proteins are not available to contribute to this film (Kiosseoglou and Sherman, 1983a,b). At low use levels, this limited availability of proteins, causes gaps in the protective films around the fat droplets and allows proteins to bridge between droplets to form loosely associated or flocculated droplets (Tornberg et al., 1990). Optimum use of yolk proteins in emulsions is possible by adding salt to disperse granules in food emulsions at pH 7±1 and by limiting salt content of acidic food emulsions (Bringe et al., 1996).

Protect polyunsaturated fatty acids from oxidation

Emulsifiers can protect fat droplets from oxidizing. To illustrate this antioxidant property, we emulsified flax seed oil, containing 57% linolenic acid, using extracted yolk or soy lecithin and compared the stability of the emulsified oil with a non-emulsified control. The soy lecithin and extracted yolk limited oxidation of flax seed oil, stored for 18 days at 40°C, as determined by the lower formation of propanal and hexanal, compared to the control (Fig. 16). Propanal is the main oxidation product of alpha-linolenic acid (Frankel, 1993).

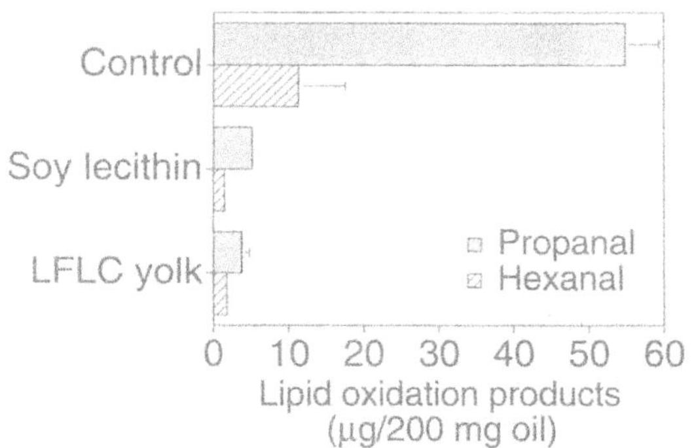

Figure 16. Oxidation of flaxseed oil as related to the addition of soy lecithin or LFLC yolk. Suspensions of commercial flaxseed oil (20%, w/w; Source Naturals Inc., Scotts Valley, CA) in salt-water (3% NaCl in water phase) were sonicated with and without the addition of soy lecithin (2%) or LFLC yolk (1.4% yolk protein, 0.6% yolk phospholipid) at pH 6.6 and 0.02% NaN3, and portions containing 200 mg oil, were sealed in 20 mL headspace vials and placed in an autoshaker (300 rpm) at 40°C for 18 days. There was less than 0.5 µg hexanal and propanal in the samples at the start of the storage study. Propanal and hexanal were quantified using a Hewlett-Packard 5970 headspace gas chromatography - mass spectrometer, a Hewlet-Packard 7694 headspace sampler, and a one-point external calibration method. The vials were equilibrated 10 min at 80°C before autosampling.

Aggregate to form fine-structured gels

The gel forming properties of egg yolk are responsible for the smooth, moist textures of foods such as baked-custard and scrambled eggs. However, gelation of egg yolk or whole eggs during frozen storage is undesirable, limiting our ability to use frozen storage to preserve these products . The formations of these desired and undesired gels are caused by the unique aggregation reactions of yolk lipoproteins, based on the special structures of yolk lipoproteins. The lipoproteins largely responsible for these reactions are the LDL, and the components of LDL which aggregate are the vitellenins (Burley and Vadehra, 1989).

LDL phospholipids are responsible for maintaining the solubility of the vitellenins. In support of this idea, vitellenins can be obtained by extracted the lipids of LDL using organic solvents, but the resulting aggregated proteins are insoluble in water (Mizutani and Nakamura, 1987). Vitellenins also aggregate when yolk is treated with phospholipase-C, which removes the polar moiety of lecithin (Mahadevan et al., 1969).

Methods used to limit freezing-induced aggregation of vitellenins involve the use of additives (sugar or salt) to increase the amount of unfrozen water and thereby prevent changes in the physical state of the phospholipids, stabilizing phospholipid-protein complexes (Smith and Back, 1975; Wakamatu et al., 1983). Another approach which limits freezing-induced aggregation of LDL is to increase the polarity of the phospholipids, using phospholipase A, which removes the fatty acids of phospholipids in the sn-2 position (Feeney et al., 1954). This approach is also effective for making LDL more resistant to heat-induced aggregation, which is especially valuable when LDL is used as a food emulsifying agent (Shenk, 1991; Carrell et al., 1992).

Yolk-lipoproteins are valuable gelling agents in foods. LDL particles are unique in their ability to form fine structured gels over a wide pH range (4-9) likely in-part, because of their content of protein-bound carbohydrate. In contrast, egg albumin proteins form weak gels consisting of large aggregates near the isoelectric pH of the proteins (Nakamura et al., 1982).

There are synergistic effects when gels are formed from mixtures of egg white and yolk. As explained by Woodward (1990), the gel forming properties of egg white are enhanced in whole egg, because the binding of yolk iron by ovotransferrin increases the denaturation temperature of ovotransferrin (from 61°C to 78°C), limiting its tendency to aggregate. Another synergistic effect of combining yolk and white is that egg albumin proteins re-enforce the fine LDL gel by coating the strands and particles of the LDL gel (Kojima and Nakamura, 1985). The bulk of the LDL particles, which are 87% fat, is not present in extracted yolk, so these textural benefits of LDL are not expected for extracted yolk.

Egg white has a significant role in egg gels made from native or LFLC egg yolk and is responsible for the higher water holding capacity of eggs at higher pH in the range 5.5 to 9.0. The higher the pH, the less serum which can be expressed from whole egg gels (Feiser and Cotterill, 1982; Woodward, 1990. This observation was also made after preparing low fat scrambled eggs and determining the amount of serum expressible from the eggs prepared at different pH. The cooked eggs prepared at pH 7 - 8 held the water phase better than the eggs at lower pH (Fig. 17). Low fat eggs expressed more water than the native eggs because they contain more water, as a result of removing fat. The LFLC eggs contained yolk lipoproteins which were not present in the egg substitute, but these lipoproteins did not improve the water holding capacity of the low fat eggs. As a texturizing agent, extracted yolk retains it ability to tenderize bakery products, but not its ability to form high water holding egg gels. The main benefit of extracted yolk in egg gels is flavor.

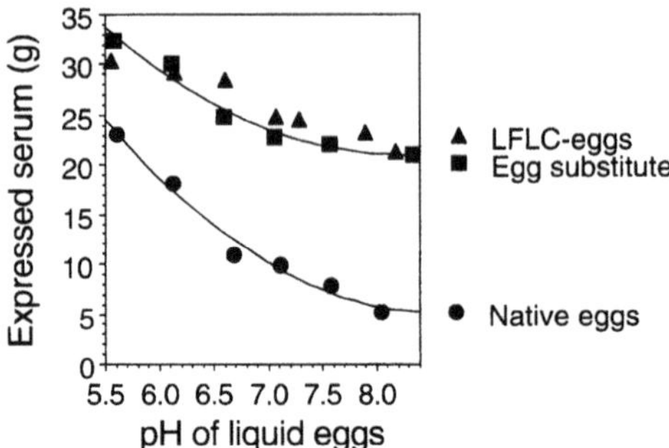

Figure 17. Expressed serum from scrambled eggs as related to pH of liquid eggs. The pH of egg samples were adjusted using 1N citric acid or 1N NaOH. Scrambled eggs were cooked in a frying pan to about 80°C, cooled to room temperature; 80 g portions were centrifuged at 25,000 x g for 30 min and the serum was weighed. The egg substitute is a leading refrigerated commercial brand (99% egg white). LFLC yolk eggs were made using egg white, rehydrated LFLC yolk, water, salt, hydroxypropylmethylcellulose, locust bean gum, citric acid or NaOH, turmeric, vitamin A palmitate, and vitamin D3; the yolk protein concentration was the same as that of native egg, 5.26%.

Bind ions

The granule phosphoproteins and phosvitin are good ion binders. Yolk granules containing phosvitin, help to prevent lipid oxidation in food systems containing iron because phosvitin has extra capacity to bind iron in the foods (Taborsky, 1963). Phosvitin is limited in iron binding by not being effective at removing iron from hemin (in meats) (Lu and Baker, 1986) and by reduced capacity to bind iron below pH 6 (Hegenauer et al., 1979). There is some free iron in egg yolk which can react with hydrogen sulfide from the albumen to form a greenish-black layer on the surface of eggs heated in the shell at temperatures above about 80°C (Burley and Vadehra, 1979). The free iron ions can also participate in another color reaction with egg white protein which is described below.

Adsorb and scatter light

Extracted yolk impacts the color of foods because it retains some egg yolk pigments which are valuable for imparting rich appearances to low-fat foods such as ice cream, pound cake and mayonnaise (Bringe and Cheng, 1985). Egg yolk granules also scatter light causing a whitening effect.

176

Egg yolk iron can also cause a color reaction which can be seen by aging freshly cracked eggs for a couple of hours (Fig. 18). The formation of the red color is the result of the binding of yolk-iron by ovotransferrin in egg white. The extent of complex and color formation in native and LFLC eggs depends of pH (Fig. 19). The iron-ovotransferrin complex formed in liquid eggs at pH values above 6. This pH effect was greatly reduced in the LFLC eggs prepared without added egg white. There is egg white contaminant in the LFLC yolk (3.7% of LFLC yolk solids is egg white solids), accounting for the redness in this product. Ovotransferrin complexes with iron via the imidazole nitrogen of a histidine group. Histidine has a pK near 6, so removing the hydrogen ion from histidine nitrogen by increasing pH, allows the complex to form. Pasteurization at 62°C had no effect on liquid egg color. Cooking (80°C) denatures ovotransferrin, destroying the colored complex. Consequently there is a relatively small effect of pH on the color of cooked eggs compared to uncooked eggs (Fig. 19). Non-enzymic browning of the eggs during cooking was favored with increases in pH above 6, accounting for the small increase in color in the cooked eggs above pH 6 (Labuza and Baisier, 1992).

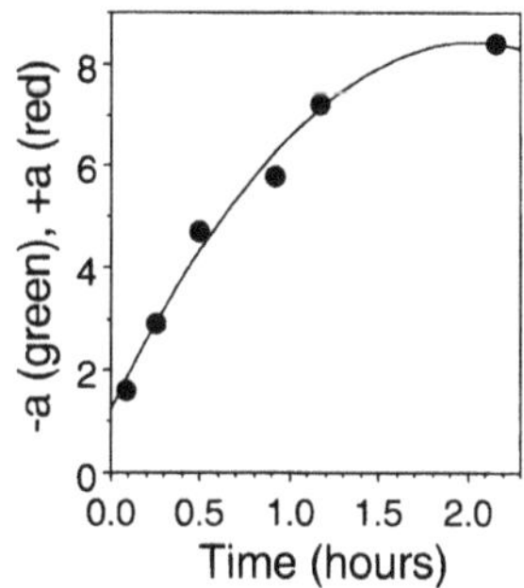

Figure 18. Redness of freshly cracked eggs as a function of mixing time. Fresh eggs (8) were cracked in a stephan bowl and mixed under vacuum (500 rpm, 3 min). The color of the liquid egg at room temperature was measured with time using a HunterLab (Reston, Va.) colorimeter.

Affect body structures and processes

We are just beginning to understand the functions of yolk-lipoproteins as nutrients, where they affect our body structures and biological processes. Yolk phospholipids are valued as a source of choline which is an important nutrient in brain development, liver function and cancer prevention (Zeisel, 1992). The phospholipids can also be a balanced source of docosahexanenoic acid and arachidonic acid, particularly valued for proper eye and brain development in infants (Simopoulos and Salem, 1992). Yolk proteins have a good balance of amino acids which humans cannot synthesize (Burley and Vadehra, 1989). Yolk glycoproteins (Koketsu et al., 1995) and egg yolk Immunoglobulin Y (Shimizu et al., 1988) provide local passive protection of the gastro-intestinal tract of infants.

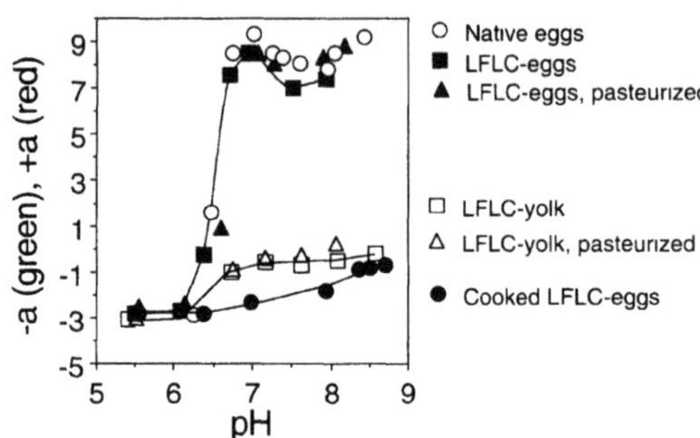

Figure 19. Redness of liquid and cooked eggs as related to pH. Cracked and mixed native eggs were aged 1 day. LFLC eggs were made with and without egg white and with and without pasteurization (3.5 min at 62°C in Stephan cooker). Yolk protein concentration was constant, 5.26%. LFLC eggs were cooked in a frying pan to about 80°C, cooled to room temperature and chopped into smaller pieces with a spatula. The cooked eggs were centrifuged and the pH of the serum was determined. Color of liquid and cooked eggs were measured using a HunterLab colorimeter.

CONCLUSION

The lipoproteins of low-fat, low-cholesterol yolk, have valuable properties which are similar to native yolk, superior to other ingredients, and are especially valuable for improving the quality of low-fat foods.

ACKNOWLEDGEMENTS

The author is also grateful to Monsanto Company management, especially Deane Clark and Suseelan Pookote, for their support and encouragement; and wishes to thank David Howard to conducting physical property experiments; Bert Chi for gas chromatography analyses; Susan Colburn for electron microscopy analyses; Karen Ferraro and Jeanette Ziegler for conducting sensory tests; Kernon Gibes for providing statistical support and Anne Hendrickson for conducting literature searches.

REFERENCES

Awazuhara, H. and Nakamura, R. 1986. Comparison of the foaming properties between egg yolk and albumen. *Lebensm.-Wiss. u. -Technol.* 19: 180-183.

Bringe, N.A. and Cheng, J. 1995. Low-fat, low-cholesterol egg yolk in food applications. *Food Technology* 49: 94-96, 98, 100, 102, 104, 106.

Bringe, N.A., Howard, D.B. and Clark, D.R. 1996. Emulsifying properties of low-fat, low-cholesterol egg yolk prepared by supercritical CO_2 extraction. *J. Food Sci.* 61:19-23, 43.

Boggs, M.M., Dutton, H.J., Edwards, B.G. and Fevold, L. 1946. Relation of lipide and salt-water fluorescence values to palatability. *Ind. Eng. Chem.* 38: 1082-1084.

Buera, M.P. and Karel, M. 1995. Effect of physical changes on the rates of nonenzymic browning and related reactions. *Food Chemistry* 52: 167-173.

Burley, R.W. and Cook, W.H. 1961. Isolation and composition of avian egg yolk granules and their constituent alpha- and beta-lipovitellins. *Can. J. Biochem. Physiol.* 39:1295-1307.

Burley, R.W. and Vadehra, D.V. 1989. *The Avian Egg. Chemistry and Biology.* John Wiley & Sons, New York.

Carrell, R.S., van Dijk, W., Goodard, M.R., and Hayes, J.B. 1992. Food product. Thomas J. Lipton Co., Division of Conopco, Inc., Englewood Cliffs, N.J. U.S. Patent 5,082,674, January 21.

Causeret, D., Matringe, E., and Lorient, D. 1991. Ionic strength and pH effects on composition and microstructure of yolk granules. *J. Food Sci.* 56: 1532-1536.

Causeret, D., Matringe, E., and Lorient, D. 1992. Mineral cations affect microstructure of egg yolk granules. *J. Food Sci.* 57:1323-1326.

Chang, C.M., Powrie, W.D. and Fennema, O. 1977. Microstructure of egg yolk. *J. Food Sci.* 42: 1193-1200.

Cotterill, O.J. and Glauert, J.L. 1979. Nutrient values for shell, liquid/frozen, and dehydrated eggs derived by linear regression analysis and conversion factors. *Poult. Sci.* 58:131-134.

Dyer-Hurdon, J.N. and Nnanna, I.A. Cholesterol content and functionality of plasma and granules fractionated from egg yolk. *J. Food Sci.* 58: 1277-1281.

Feeney, R.E., MacDonell, L.R. and Franenkel-Conrat, H. 1954. Effects of crotoxin (lecithinase A) on egg yolk and yolk constituents. *J. Biol. Chem.* 229: 130-140.

Feeney, R.E. 1989. Food technology and polar exploration. *Food Technology* 43(5): 70-82.

Frankel, E.N. 1993. Formation of headspace volatiles by thermal decomposition of oxidized fish oils vs. oxidized vegetable oils. *J. Am. Oil Chem. Soc.* 8: 767-772.

Froning, G.W., Wehling, R.L., Cuppett, S.L., Pierce, M.M., Niemann, L. and Siekman, D.K. 1990. Extraction of cholesterol and other lipids from dried egg yolk using supercritical carbon dioxide. *J. Food Sci.* 55(1): 95-98.

Gardner, F.A., Beck, M.L., and Denton, J.H. 1982. Functional quality comparison of whole egg and selected egg substitute products. *Poultry Sci.* 61:75-78.

Graham and Kamat, 1977. The role of egg yolk lipoproteins in fatless sponge cake making. *J. Sci. Food Agric.* 28: 34-40.

Griffin, H.D. 1992. Manipulation of egg yolk cholesterol: a physiologist's view. *World's Poultry J.* 48:101-112.

Hahn, P.W. 1993. Low-fat cereal-grain food composition. The Pillsbury Company, Minneapolis, Minn. U.S. Patent 5,262,187, November 16.

Hegenauer, J., Saltman, P., and Nace, G. 1979. Iron (III)-phosphoprotein chelates: stoichiometric equilibrium constant for interaction of iron (III) and phosphorylserine residues of phosvitin and casein. *Biochemistry* 18:3865-3874.

Jost, R.; Dannenberg, F.; Rosset, J. Heat-set gels based on oil/water emulsions: an application of whey protein functionality. *Food Microstructure* 8: 23-28.

Junge, R.C.; Hoseney, R.C. 1981. A mechanism by which shortening and certain surfactants improve loaf volume in bread. *Cereal Chem.* 58(5): 408-412.

Kilara, A. and Shahani, K.M. 1973. Removal of glucose from eggs: a review. *J. Milk Food Technology* 36: 509-514.

Kiosseoglou, V.D. and Sherman, P. 1983a. The influence of egg yolk lipoproteins on the rheology and stability of O/W emulsions and mayonnaise. 2. Interfacial tension-time behaviour of egg yolk lipoproteins at the groundnut oil-water interface. *Colloid and Polymer Sci.* 261: 502-507.

Kiosseoglou, V.D. and Sherman, P. 1983b. The influence of egg yolk lipoproteins on the rheology and stability of O/W emulsions and mayonnaise. 3. The viscoelastic properties of egg yolk films at the groundnut oil-water interface. *Colloid and Polymer Sci.* 261: 520-526.

Kline, L., Gegg, J.E., and Sonoda, T.T. 1951a. Role of glucose in the storage deterioration of whole egg powder. II. A browning reaction involving glucose and cephalin in dried whole eggs. *Food Technology* 5: 181-187.

Kline, L., Hanson, H.L., Sonoda, T.T., Gegg, J.E., Feeney, R.E. and Lineweaver, H. 1951b. Role of glucose in the storage deterioration of whole egg powder. III. Effect of glucose removal before drying on organolepic, baking, and chemical changes. *Food Technology* 5: 323-331.

Kline, L., Sugihara, T.F. and Meehan, J.J. 1964. Properties of yolk-containing solids with added carbohydrates. *J. Food Sci.* 29:693-709.

Koketsu, M., Nitoda, T., Juneja, L.R., Kim, M., Kashimura, N., and Yamamoto, T. 1995. Sialyloligosaccharides from egg yolk as an inhibitor of rotaviral infection. *J. Agric. Food Chem.* 43:858-861.

Kojima, E. and Nakamura, R. 1985. Heat gelling properties of hen's egg yolk low density lipoprotein (LDL) in the presence of other protein. *J. Food Sci.* 50: 63-66.

Labuza, T.P., Baisier, W.M. 1992. The kinetics of nonenzymatic browning. Chptr 14 in *Physical Chemistry of Foods*, eds. H.G. Schwartzberg and R.W. Hartel, pp. 595-649. Marcel Dekker, Inc., New York.

Le Meste, M. and Davidou, S. 1995. Lipid-protein interactions in foods. Chptr 8 in *Ingredient Interactions - Effects on Food Quality*, ed A.G. Gaonkar, pp. 235-268. Marcel Dekker, Inc., New York.

Lu, C-L. and Baker, R.C. 1986. Characteristics of egg yolk phosvitin as an antioxidant for inhibiting metal-catalyzed phospholipid oxidations. *Poult. Sci.* 65: 2065-2070.

Mahadevan, S., Satyanarayana, T., and Kumar, S.A. 1969. Phsico-chemical studies on the gelation of hen's egg yolk. *J. Agr. Food Chem.* 17(4): 767771.

Nakamura, R., Fukano, T. and Taniguchi, M. 1982. Heat-induced gelation of hen's egg yolk low density lipoprotein (LDL) dispersion. *J. Food Sci.* 47: 1449-1453.

Pike, O.A. and Peng, I.C. 1985. Stability of shell egg and liquid yolk to lipid oxidation. *Poult. Sci.* 64: 1470-1475.

Pike, O.A. and Peng, I.C. 1988a. Effect of protein disruption by denaturation and hydrolysis on egg yolk lipid oxidation. *J. Food Sci.* 53(2): 428-431.

Pike, O.A. and Peng, I.C. 1988b. Influence of pH on egg yolk lipid oxidation. *J. Food Sci.* 53(4): 1245-1246.

Powrie, W.D. 1977. Chemistry of eggs and egg products. Chptr. 6 in *Egg Science and Technology*, eds. W.J. Stadelman and O.J. Cotteril, Avi Pub. Co. Inc. Westport, Connecticut.

Satyanarayana Rao, T.S. and Murali, H.S. 1988. Effect of moisture on solubility, free fatty acid content and beta-carotene of spray-dried, freeze-dried and foam-mat-dried egg powder during storage. *J. Fd. Sci. Technol.* 25(6): 368-370.

Schultz, J.R., Snyder, H.E., and Forsythe, R.H. 1968. Co-dried carbohydrates effect on the performance of egg yolk solids. *J. Food Sci.* 33: 507-513.

Shenk, B. 1991. Process for the preparation of a water and oil emulsion. Van den Bergh Foods Company, Division of Conopco, Inc., New York, N.Y. U.S. Patent 5,028,447, July 2.

Shimizu, M., Fitzsimmons, R.C., Nakai, S. 1988. Anit-E. coli immunoglobulin Y isolated from egg yolk of immunized chickens as a potential food ingredient. *J. Food Sci.* 53:1360-1366.

Simopoulos, A.P., and Salem, N. 1992. Egg yolk as a source of long-chain polyunsaturated fatty acids in infant feeding. *Am. J. Clin. Nutr.* 55:411-414.

Slade, L.; Levine, H.; Finley, J.W. 1989. Protein-water interactions: water as a plasticizer of gluten and other protein polymers. Chapter 2 in *Protein Quality and the Effects of Processing*, ed. R. D. Phillips and J.W. Finley, pp. 9-124. Marcel Dekker, Inc., New York.

Smith, M.B. and Back, J.F. 1975. Thermal transitions in the low-density lipoprotein and lipids of the egg yolk of hens. *Biochim. Biophys. Acta* 388: 203-212.

Taborsky, G. 1963. Interaction between phosvitin and iron and its effect on a rearrangement of phosvitin structure. *Biochemistry* 2:226-271.

Tarladgis, B.G., Watts, B.M., and Younathan, M.T. 1960. A distillation method for the quantitative determination of malonaldehyde in rancid foods. *J. Am. Oil Chem. Soc.* 37: 44-48.

Tornberg, E., Olsson, A., and Persson, K. 1990. The structural and interfacial properties of food proteins in relation to their function in emulsions. Ch. 7 in *Food Emulsions*, 2nd edition, K. Larsson and S.E. Friberg (Eds.), pp. 247-322. Marcel Dekker, New York.

Van Elswyk, M.E., Sams, A.R., and Hargis, P.S. 1992. Composition, functionality, and sensory evaluation of eggs from hens fed dietary menhaden oil. *J. Food Sci.* 57(2): 342-344, 349.

Van Elswyk, M.E., Dawson, P.L., and Sams, A.R. 1995. Dietary menhaden oil influences sensory characteristics and headspace volatiles of shell eggs. *J. Food Sci.* 60(1): 85-89.

Verstrate, J.A. 1980. Chemical and organoleptic characterization of flavor changes during storage of hard-cooked egg albumen and yolk. Ph.D. thesis, Washington State University.

Wakamatu, T., Sato, Y. and Saito, Y. 1983. On sodium chloride action in the gelation process of low density lipoprotein (LDL) from hen egg yolk. *J. Food Sci.* 48: 507-512, 516.

Warren, M.W., Ball, H.R.Jr., Froning, G.W. and Davis, D.R. 1991. Lipid composition of hexane and supercritical carbon dioxide reduced cholesterol dried egg yolk. *Poult. Sci.* 70:1991-1997.

Warren, M.W., larick, D.K., and Ball, H.R. Jr. 1995. Volatiles and sensory characteristics of cooked egg yolk, white and their combinations. *J. Food Sci.* 60:79-84, 97.

Wasan D.T. 1992. Interfacial transport processes and rheology: structure and dynamics of thin liquid films. *Chem. Eng. Ed.*, Spring issue, 104-112.

Woodward, S.A. 1990. Egg protein gels. Chpt. 5 in *Food Gels*, ed. P. Harris, pp. 175-199. Elsevier Applied Science. New York.

Woodward, S.A., and Cotterill, O.J. 1987. Texture and microstructure of cooked whole egg yolks and heat-formed gels of stirred egg yolk. *J. Food Sci.* 52:63-67.

Yamamoto, Y., Sogo, N., Iwao, R., and Miyamoto, T. 1990. Antioxidant effect of egg yolk on linoleate in emulsions. *Agric. Biol. Chem.* 54(12): 3099-3104.

Zeisel, S.H. 1992. Choline: an important nutrient in brain development, liver function and carcinogenesis. *J. Am. Coll. Nutr.* 11(5):473-481.

INTERACTIONS BETWEEN DIETARY PROTEINS AND THE HUMAN SYSTEM: IMPLICATIONS FOR ORAL TOLERANCE AND FOOD-RELATED DISEASES

William E. Barbeau

Department of Human Nutrition and Foods
Virginia Polytechnic Institute and State University
Blacksburg, Virginia 24061

ABSTRACT

Mounting evidence suggests that there are a number of important, but poorly understand, interactions between dietary proteins and the human immune system. The usual response of the human immune system to dietary proteins seems to be that of oral tolerance, a phenomenon involving up-regulation of protective gut localized immune mechanisms and down-regulation of potentially harmful systemic immunity to the protein in question. Abrogation of oral tolerance may play an important role in the development of food allergies and food enteropathies. Immune mechanisms underlying oral tolerance are therefore discussed in light of current understanding of such food-related diseases as IgE mediated food allergies and gluten sensitive enteropathy. Possible development of oral vaccines to immune-related diseases like multiple sclerosis and rheumatoid arthritis is also discussed.

INTRODUCTION

We humans are exposed each day of our lives to a host of different substances. Some of these substances are free amino acids or substances comprised of amino acids, peptides and polypeptides (proteins). Human exposure to proteinaceous substances occurs in a variety of ways e.g. skin and mucous membrane exposure, airway and lung exposure to aerosols, pollen etc. The best-described interactions between humans and proteinaceous substances in the environment, however, involve those proteins that are ingested. The essential amino acids that humans need as as building blocks for de novo protein synthesis are (except in the case of parenteral feeding of free amino acids) all derived from ingested or dietary proteins (Munro and Crim, 1980). A second, reasonably well-described interaction between humans and ingested proteins, involves a small minority of dietary proteins that have been labeled as antinutrients,

Food Proteins and Lipids
Edited by Damodaran, Plenum Press, New York, 1997

Table 1: Serum Antibody Levels After Oral Immunization.

Oral Immunization	Intravenous[1] Challenge	Anti-BGG[2]			Anti-Gliadin2		
		IgA	IgG	IgM	IgA	IgG	IgM
BGG, 14 weeks	BGG	3.58*	0.32	0.52*	0.31	-0.21	-0.08
Gliadin, 14 weeks	Gliadin	0.57	0.01	-0.06	1.95*	0.50	1.22*
BGG, 14 weeks	Gliadin	3.34*	0.29	-0.28	0.09	1.18*	-0.12
Gliadin, 14 weeks	BGG	0.37	1.25*	0.24	3.82*	1.24*	0.15

[1]Three consecutive daily intravenous injections of the indicated protein after 14 weeks of oral immunization.
[2]Data are number of doubling dilutions in ELISA antibody titer of indicated class and specificity relative to normal mouse serum.
*Significantly different (p<0.01) from control mouse serum. [Reproduced from Emancipator and Lamm (1988), with permission from S. Karger AG, Basel.]

e.g. soy trypsin inhibitors, black bean hemagglutins etc., because of their interference with nutrient digested and/or absorption (Liener, 1980).

Dietary proteins also come in contact with human immune system. Interactions between the human immune system and dietary proteins appear to be multi-faceted, complex and of enormous importance to human health and wellbeing. Normal interactions between dietary proteins and the human immune system are grouped together under the heading of a phenomenon known as oral tolerance. Most individuals have developed, or will develop, some degree of oral tolerance to at least ninety-nine percent of the proteins in their diet. Oral tolerance begins to develop soon after birth. However, newborns continue to be sensitive to dietary proteins for a number of months, or longer, until nearly full maturation of the infant's digestive and immune systems occurs (Strobel and Ferguson, 1984).

DEFINITION AND MECHANISMS OF ORAL TOLERANCE

One of the first well-controlled investigations of oral tolerance was conducted near the turn of the century by H. Gideon Wells. Wells knew that repeated injections of protein antigens caused potentially fatal, anaphylactic reactions in guinea pigs. What would happen, he wondered, if the guinea pigs were fed the protein antigens prior to receiving injections? The results of his experiments on egg white proteins are summarized in his 1911 paper (Wells, 1911), "Guinea-pigs fed upon a certain protein are at first rendered sensitive to this protein. After some time, however, if the feeding is continued they become less sensitive, until they reach an immune or refractory condition so that they do not react to two spaced injections of the fed protein.". Wells had discovered the phenomenon of oral tolerance.

Immunogists are just beginning to understand this phenomenon at thecellular and molecular level. As indicated in Figure 1, at least two immunoregulatory mechanisms are believed to be involved in development of oral tolerance: up-regulation of secretory IgA antibody production, promoting clearance of ingested antigen (protein), and suppression of systemic immune responses to the same antigen. To illicit an immune response, food allergens must first be presented to T lymphocytes. Unhydrolyzed portions of food proteins cross the gut lumen border into the small intestinal epithelium where they are endocytosed by macrophages (or alternatively by enterocytes) and after further proteolytic processing are resented as antigens to helper (T4) and suppressor (T8) T lymphocytes. Helper T cells are instrumental in up-regulating antigen-specific IgA synthesis while suppressor T cells help to abolish antigen-specific delayed type hypersensitivity (DTH) reactions

Up-regulation of IgA production, in response to ingested of food antigens, is illustrated in a study by Emancipator and Lamm (1988). These investigators found that mice orally immunized to a specific antigen (bovine gamma globulin (BGG) or gliadin) developed significantly elevated titers of anti- IgA antibodies to the fed antigen (see Table 1); whereas, intravenous challenge produced significantly elevated titers of anti-IgG and anti-IgM antibodies to innoculated antigens. IgA does not fix complement as well as IgG and IgM antibodies, and is not nearly as potent at mediating antibody-dependent cytotoxicity (Emancipator and Lamm, 1988). However, the translocation rate of IgA antibodies into the gut lumen easily exceeds that of IgG and IgM classes of antibodies (Brandtzaeg, 1989). Suppression of delayed hypersensitivity reactions to ingested antigens implies that these antigens are prevented from entering the bloodstream. Dimer and polymeric forms of IgA may be uniquely equipped at sealing mucosal surfaces and preventing ingested antigens from penetrating mucosal barriers.

There is extensive evidence that T cells also play a role in establishment of oral tolerance. Kagnoff (1978) found that he could prevent delayed-type hypersensitivity reactions in mice by two week feeding of the appropriate antigen (sheep erythrocytes (SRBC) or horse erythrocytes (HRBC)) (see Table 2). Furthermore he found that transfer of suppressor T cells

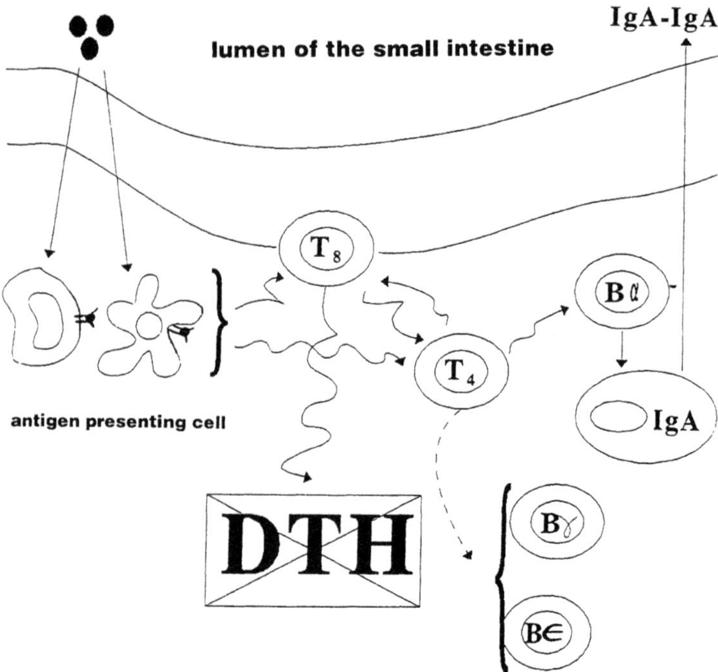

Figure 1: Hypothetical scheme for induction of oral tolerance. [Adapted from Brandtzaeg (1989) with permission from Springer-Verlag.]

from the spleens of SRBC-fed mice to normal mice (not fed SRBC) reduced footpad delayed hypersensitivity reactions to normal (results not shown). These results suggest that oral tolerance is mediated at least in part by antigen-specific suppressor T cells.

Table 2: Specificity of Diminished Delayed Type Hypersensitivity (DTH) Reactions in erythrocyte-fed Mice[1].

	DTH Reactions to	
Antigen fed	SRBC	HRBC
None	65.5±3.3	46.1±4.8
SRBC	2.1±1.3	34.0±0.9
HRBC	58.0±5.1	1.0±0.4

[1]Mice fed sheep (SRBC) or horse (HRBC) erythrocytes were sensitized with SRBC or HRBC intravenously and footpad challenged with SRBC or HRBC. DTH reactions are expressed as the percent in footpad thickness ± SE 24 hours after foodpad challenge.
[Reprinted with permission from Kagnoff (1978),
©1978 The American Association of Immunologists.]

The exact mechanisms for induction of suppressor T cells involved in oral tolerance remain unclear. The original proposal of a number of investigators (Ngan and Kind, 1978;

Richman et al., 1981), that Peyer's patches were the sites of clonal selection and expansion of these cells, has been discretided (Enders et al., 1986). Oral tolerance was still induced in rats even after surgical removal of Peyer's patches. This finding has led to the speculation that thymus-derived intraepithelial lymphocytes (IELS) and T lymphocytes in the lamina propria might be involved in regulation of immune responses to ingested antigens (Enders et al., 1986). Recently, Galliaerde et al. (1995) found that enterocytes from 2,4-dinitrochlorobenzene-fed mice inhibited hapten-primed T cells which led to the suggestion that enterocytes play a significant role in mediation of T cell responses during establishment of oral tolerance.

There is also controversy about the nature of the suppressor T cells involved in oral tolerance. Supressor T cells derived their name by virtue of the fact that these cells display certain transmembrane proteins on their cell surfaces. Induction of oral tolerance was originally thought to involve T lyphocytes displaying CD8 molecules (Lider et al., 1989) (helper T cells display CD4 molecules and supressor T cells CD8). However, Garside et al. (1995) were able to demonstrate that induction of oral tolerance ovalbumin in mice is abolished by in vitro depletion of CD4+ but not CD8+ T cells. These seemingly disparate findings may in the future be reconciled by a relatively new model for T cell identity and function espoused by Mosmann and others. One of the implications of this model is that suppressor T cells do not exist, that there are instead two kinds of helper T cells, designated THI and TH2, that can be observed to have different functions due to secretion of different patterns of lymphokines (also called cytokines) (Mosmann and Coffman, 1989).

There is obviously a lot we do not know about oral tolerance. What happens, for example, when an individual fails to develop oral tolerance to a particularly food protein or suffers a breakdown in previously acquired tolerance? It is entirely possible that many food intolerances, including food allergies and intestinal enteropathies, are due at least in part to a failure and/or a breakdown in oral tolerance.

The Aftermath Of A Breakdown In Oral Tolerance

Food Allergies. A food allergy may be defined as an abnormal or exaggerated immunologic response to a food or food component (almost always a protein) (Sampson and Metcalfe, 1991). Many more people apparently believe they have a food allergy than actually do. One survey found that a third of American women believed that at least one member of their household suffered from a food allergy (Sloan and Powers, 1986). Sampson estimates that the true prevalence of food allergies is much lower, about 5% of young children and 2% of adults (Sampson, 1992). Almost half of the food allergies in infants (prevalence rate of 2.2% in one study and 2.5% in another) are caused by cow's milk protein (Sampson, 1992). A number of studies have found that breastfeeding lowers the incidence of food allergies, in spite of the fact that breastmilk can be a vehicle for sensitizing infants to allergens (Perkin, 1990). There may be an inherited predilection towards food allergies, linked to excessive synthesis of food antigen specific IgE antibodies (Anderson, 1990).

The majority of food allergies are thought to be initiated by the binding of antigen specific IgE antibodies to mast cells and basophils leading to degranulation of these cells and release of potent chemical mediators such as histamine, proteoglycans and leukotrienes. IgE mediated immediate hypersensitivity reactions usually cause a food allergy victim to experience symptoms within minutes of ingested of a particular allergen. Actual allergic symptoms correspond to the tissue specific effects of these various mediators and may include include nausea, abdominal cramps, vomiting, diarrhea, angioedema, pruritic rashes, hives, asthma, rhinitis, laryngeal edema and in severe cases, life-threatening anaphylaxis (Barrett, 1991; Sampson and Metcalfe, 1991).

Legumes (especially peanuts and soybeans).
Crustacea: Shrimp, crab, lobster, crayfish.
Milk: Including cow's milk and goat milk.
Eggs from all avian species.
Tree nuts: Almonds, walnuts, brazil nuts, etc.
Fish: Cod, haddock, salmon, etc.
Mollusks: Clams, oysters, scallops, etc.
Wheat.

Taken with permission from Taylor (1992).

Figure 2 illustrates how IgE-mediated food allergies are believed to be triggered by degranulation of mast cells and basophils. Plasma cells synthesize and secrete antigen-specific IgE antibodies. Antigen-specific IgE molecules then migrate to and bind to mast cells or basophils. These membrane-bound IgE molecules are then bridged or cross-linked by the binding of specific food allergens. A signal is sent to the mast cell or basophil to degranulate, once a critical mass of IgE molecules have been cross-linked with food allergens (Kane et al., 1988; Taylor, 1992).

Food allergens must be of a certain molecular size and perhaps conformation to bind an bridge adjacent IgE molecules on mast cell surfaces (Taylor, 1992) Degranulation is thought to require binding of a rigid, bivalent antigen of a minimum length of about 200 angstroms (Kane et al., 1988). According to Taylor, this corresponds to a protein with a molecular weight of between 10,000 and 70,000 daltons. Proteins from a variety of sources, some plant, some animal apparently meet these criteria (see Table 3). Many food allergens are glycoproteins. (Metcalfe, 1992; Taylor, 1992). Whether the glyco or carbohydrate portion of these molecules has anything to do IgE binding, is presently unknown. Food allergens are also apparently stable to to heat and acid treatments. Furthermore, they must be resistant to peptic, typtic, chymothryptic etc. hydrolysis to reach the intestinal muscosa in a fully intact, immunogenic form (Taylor, 1992).

Food Enteropathies With An Emphasis On Gluten-Sensitive Enteropathy

Gluten sensitive enteropathy (GSE) or celiac disease is the most well-known food enteropathy. Individuals with GSE find it impossible to consume certain cereals, particularly wheat but also rye, triticale, barley, and possibly oats without incurring symptoms ranging from mild abdominal discomfort and diarrhea to joint pain, and anemia. Columnar epithelial cells may be destroyed in the small intestine causing flattening of the villi and malabsorption of such nutrients as calcium, magnesium, folic acid, vitamin B12, fat and the fat soluble vitamins (Broitman and Zamcheck, 1980; Troncone and Auricchio, 1991). Malabsorption, secondary to GSE, is so severe that it can cause permanent stunting of growth if the disease is not properly diagnosed during the first decade of life (Bosio et al., 1990). At present, the only effective treatment for GSE is to remain on a lifelong gluten-free diet.

GSE is neither very prevalent nor it is a rare disease. Approximately one out of every thousand people worldwide will eventually be diagnosed as having GSE. However, the prevalence of asymptomatic cases of the disease may, in fact, be a great deal higher (Troncone and Auricchio, 1991). In addition, some people have an intolerance for wheat which manifests itself in the development of peculiar skin lesions; this disease called dermatitis herpetiformis, is probably somehow related to GSE (Sachs et al., 1986).

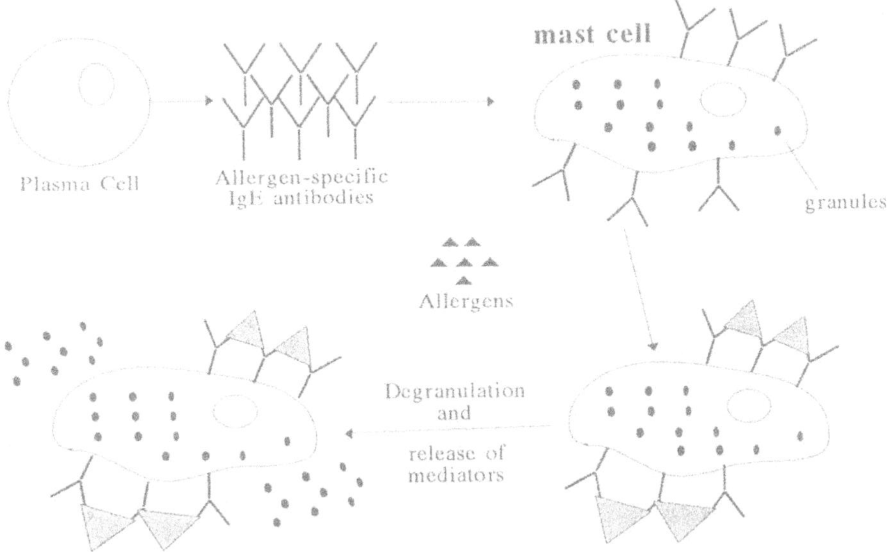

Figure 2: Proposed mechanism(s) for IgE-mediated hypersensitivity reactions. (Adapted with permission from Taylor, 1992).

The etiological agents in GSE are believe to be small peptides that are derived from the ethanol-soluble proteins (prolamins) of wheat, rye, triticale, oats and barley (Troncone and Auricchio, 1991). Cereals are not the only foodstuffs that trigger food enteropathies in humans. A number of other proteins from both plant and animal sources such as soy protein (Ament and Rubin, 1972); fish, rice and chicken protein (Victoria et al. , 1982); egg protein (lyngkaran et al., 1982); and cow's milk protein (lyngkaran et al., 1988) have also been incriminated in food intolerances and intestinal enteropathies. Since whole rather than fractionated proteins were fed in these studies, the particular protein moiety triggering intestinal enteropathy is, in these cases, unknown.

Virtually nothing is known about the etiology and pathogenesis of food enteropathies other than GSE. It was originally proposed that GSE resulted from a missing or deficient enzyme, thought to be a brush border protease or carbohydrase, that led to the accumulation of cytotoxic gliadin peptides (Cornell and Rolles, 1978; Peters et al., 1978). A subsequent hypothesis attributed small bowel damage in GSE to cytotoxicity of lectin-like gliadin peptides (Weiser and Douglas, 1976), or alternatively, to a well-known lectin, wheat germ agglutinin (Colyer et al., 1986). Skepticism about these two hypotheses led Kagnoff to suggest that GSE is caused by molecular mimicry between one or more gliadin peptides that are presumably refractory to intraluminal and intracellular proteases and a peptide sequence present in an adenovirus protein, Elb (Kagnoff et al., 1984).

The exact cause(s) of GSE remains a mystery. It seems clear however that much of the small bowel pathology in GSE is caused by the patient's own immune system (O'Farrelly and Gallagher, 1992). The likelihood of eventually developing either an assymptomatic or a full-blown case of the disease is linked to one's inheritance of chromosome 6 genes encoding proteins involved in antigen presentation (Marsh,1992). Another intriguing element of GSE pathogenesis is the presence in GSE lesions of elevated numbers of gamma delta T cell receptor (TCR) bearing T lymphocytes. Numbers of these cells remain elevated even when patients go on a gluten-free diet (Trejdosiewicz et al., 1991; Halstensen et al., 1989).

Gamma/delta T cells have been called enigmatic molecules of the human immune system (Strominger, 1989) because their origin and exact functions remain unknown. These cells bind to antigen presenting cells using a TCR composed of two proteins, a gamma and delta heterodimer. This heterodimer is different than the alpha beta TCR heterodimer expressed by most T lymphocytes. Activated gamma delta T cells preferentially home to the skin, lungs, GI tract, and spleen (Augustin et al., 1989; Bluestone and Matis, 1989), where they may respond to invading pathogens such as M. tuberculosis and M. leprae (Raulet, 1989).

We have recently advanced a new hypothesis for the etiology and pathogenesis of GSE (Barbeau et al., 1996) which, if correct, might explain why there is an influx into GSE lesions, of gamma delta T cells. There are some similarities between our hypothesis and that of Kagnoff's. We too believe that GSE is triggered by a case of mistaken identity, that there is sufficient molecular mimicry between HLA class 11 dimer bound forms of gliadin peptides and those of a particular bacterial or viral antigen to be recognized by identical gamma delta T cell subsets. Furthermore, we propose that some individuals are more susceptible to GSE than others due to inheritance of genes encoding critical HLA class 11 molecules that are involved in the antigen presentation of gliadin peptides.

According to our theory, putative gladin peptides bind distal to the centrally located peptide binding groove on HLA class 11 molecules, to N-linked oligosaccharides that extend away from the class 11 molecule (see Figure 3). Evidence suggests that bacterial and

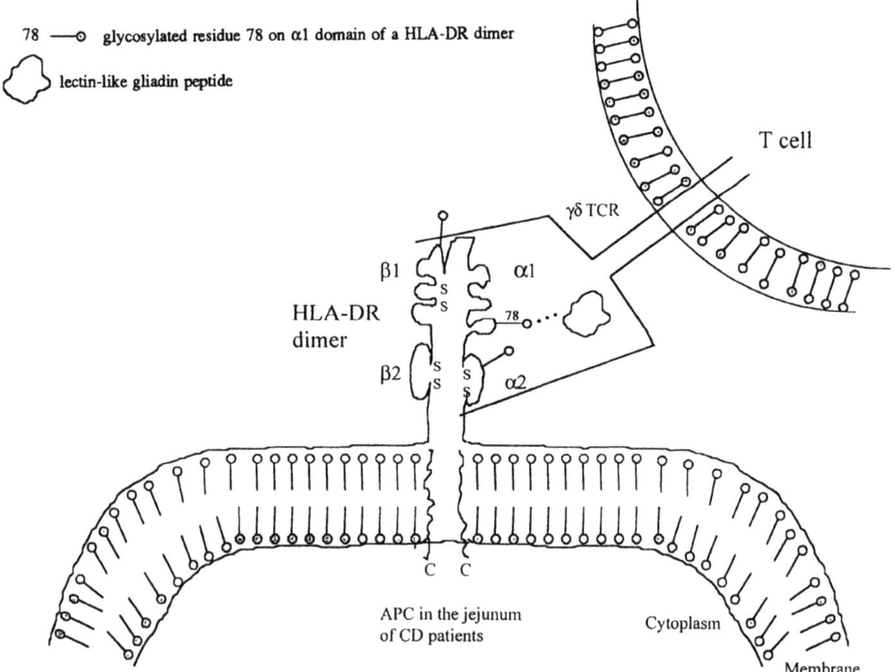

(Please note that according to Brown et al. (1993) HLA class II molecules may exist in situ as dimers)

Figure 3: A proposed mechanism for recognition and binding of gliadin peptides and CD associated HLA-DR complexes by gamma delta T cells.

viral superantigens bind in a similar fashion to HLA class 11 molecules (Dellabona et al, 1990) and that HLA class 11 bound forms of these superantigens are specifically engaged by gamma delta TCR bearing T lymphocyes (Schild et al., 1994). Gamma delta T cell engagement sets in motion a series of humoral and T cell mediated immune responses that lead to the destruction of the epithelial lining of the small intestine (see Figure 4).

Members of the general population may also bind and present putative gliadin peptides as antigens but would do so using other HLA class 11 molecules (again see Figure 4). We theorize that the putative gliadin peptides bind in this case to the centrally located peptide binding groove on HLA class 11 molecules. Gliadin peptides presented in this fashion are recognized and bound by alpha beta TCR bearing T lymphocytes. Instead of causing a self-destructive immune response, these alpha beta T lymphocytes help to induce a state of oral tolerance to ingested wheat gliadins.

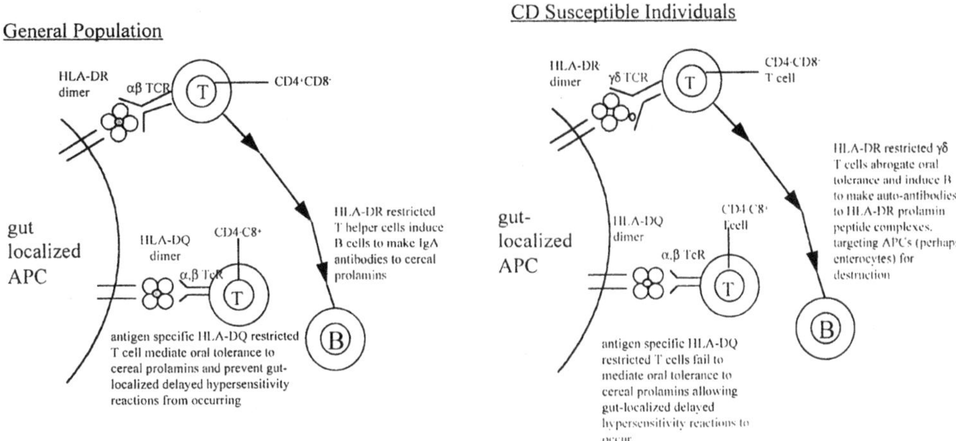

Figure 4: Possible senarios following ingestion of wheat and other cereals leading to oral tolerance in the general population and to gut-localized delayed hypersensitivity in CD susceptible individuals.

What Lies Ahead?

Greater understanding of the complexities of oral tolerance and diseases such as GSE may eventually lead to development of successful therapeutic and prophylactic measures for a host of immune-related diseases. Investigators are presently exploring the possibilities of developing oral vaccines to treat diseases such as autoimmune encephalomyelitis (EAE), multiple sclerosis, rheumatoid arthritis and juvenile onset diabetes (IDDM) (Friedman et al.,1994; Weiner et al., 1994). It may not even be necessary to identity the offending antigen (the exogenous or endogenous antigen(s) triggering the disease) to develop a vaccine. Investigators have found that, in some cases, they can suppress disease symptoms by oral administration of a bystander antigen (a non-disease inducing portion of an autoantigen) (Miller et al., 1991; Weiner et al., 1994).

REFERENCES

Ament, M.E. and Rubin, C. E. 1972. Soy protein-another cause of the flat intestinal lesion. Gastroenter. 62: .227.

Anderson J.A. 1990 Food allergy or sensitivity terminology, physiologic bases, and scope of the clinical problem. Ch. 1. In Food Allergies and Adverse Reactions, J.E. Perkin (Ed.), p. 5. Aspen Publishers, Inc., Gaithersburg, MD.

Augustin, A., Kubo, R.T., Sim, G-K., 1989. Resident pulmonary lymphocytes expressing the gamma/delta cell receptor. Nature. 340: 239.

Barbeau, W.E., Novascone, M.A., Elgert, K.D. 1996. Reappraisal of the lectin and immune hypotheses for the etiolgy of celiac disease in light of possible interactions between gliadin peptides and HLA class 11 molecules. Gut., in review.

Barrett, K.E. 1991. Mast cells, basophils and immunoglobulin E. Ch. 2 In Food Allergy:Adverse Reactions to Foods and Food Additives, D.D. Metcalfe, H.A. Sampson and R.A. Simon (Ed.), p. 28-29. Blackwell Scientific Publications, Boston, MA.

Bluestone J.A. and Matis, L.A. 1989. TCR gamma delta cells - minor redundant T cell subset or specialized immune system component? J. Immunol. 142: 1785.

Bosio, L., Barera, G., Mistura, L., Sassi, G., Bianchi, C. 1990. Growth acceleration and final height after treatment for delayed diagnosis of celiac disease. J. Pediatr. Gastroenter. Nutr. 11:324.

Brandtzaeg, P. 1989. Overview of the mucosal immune system. Curr. Topics Microbiol. Immunol. 146: 13.

Broitman, S.A. and Zamcheck, N. 1980. Nutrition in diseases of the intestines. Ch. 31B. In Modern Nutrition in Health and Disease, 6th ed. , R.S. Goodhart and M.E. Shils (Ed.), p. 922. Lea & Febiger, Philadelphia, PA.

Brown, J.H., Jardetzky, T. S., Gorga, J.C., Stern, L.J., Urban, R.G., Strominger, J.L., Wiley, D.C. 1993. Three-dimensional structure of the human class 11 histocompatibility antigen HLA-DRI. Nature. 364: 33.

Colyer, J., Farthing M.J.G., Kumar, P.J., Clark, M.L., Ohannesian, A.D., Waldron, N.M. 1986. Reappraisal of the 'lectin hypothesis' in the aetiopathogenesis of coeliac disease. Clin Sci. 71: 105.

Cornell, H.J. and Rolles, C.J. 1978. Further evidence of a primary mucosal defect in coeliac disease. Gut 19: 253.

Dellabona , P., Peccoud, J., Kappler, J., Marrack, P., Benoist, C., Mathis, D. 1990. Superantigens interact with MHC class 11 molecules outside of the antigen groove. Cell. 62: 1115.

Emancipator, S.N. and Lamm, M.E. 1988. Oral tolerance as a protective mechanism against hypersensitivity disease. Monogr. Allergy. 24: 244.

Enders, G., Gottwald, T., Brendel, W. 1986. Induction of oral tolerance in rats without Peyer's patches. Immunol. 58: 311.

Friedman, A., Al-Sabbagh, A., Santos, L.M.B., Fishman-Lobell, J., Polanski, M., Das, M.P., Khoury, S.J., Weiner, H.L. 1994. Oral tolerance: a biologically relevant pathway to generate peripheral tolerance against external and self antigens. Ch. 10 In Chemical Immunology. Mechanisms of Immune Regulation., R.D. Granstein (Ed.), Vol. 58, p. 259. Karger, Basel, Switzerland.

Galliaerde, V., Desvignes, C., Peyron, E., Kaiserlian, D. 1995. Oral tolerance to haptens: intestinal epithelial cells from 2,4-dinitrochlorobenzene-fed mice inhibit hapten-specific T cell activation in vitro. Eur. J. Immunol. 25: 1385.

Garside, P., Steel, M., Liew, F.Y., Mowat, A.M. 1995. CD4+ but not CD8+ T cells are required for the induction of oral tolerance. Internat. Immunol. 7: 501. Halstensen, T.S., Scott, H., Brandtzaeg, P. 1989. Intraepithelial T cells of the TcR gamma/delta+ CD8- and Vdeltal/jdeltal+ phenotypes are increased in coeliac disease. Scand. J. Immunol. 30: 665.

lyngkaran, N., Abidin, Z., Meng, L.L., Yadav, M. 1982. Egg protein-induced villous atrophy. J. Pediatr. Gastroenter. Nutr. 1: 29.

lyngkaran, N., Yadav, M., Boey, C.G., Lam K.L. 1988. Severity and extent of upper small bowel mucosal damage in cow's milk protein-sensitive enteropathy. J. Pediatr. Gastroenter. Nutr. 7: 667.

Kagnoff, M.F. 1978. effects of antigen-feeding on intestinal and systemic immune responses. 11. Suppression of delayed-type hypersensitivity reactions. J. Immunol. 120:1509.

Kagnoff, M.F., Austin, R.K., Hubert, J.J., Bernardin, J.E., Karsarda, D.D. 1984. Possible role for a human adenovirus in the pathogenesis of celiac disease. J. Exp. Med. 160: 1544.

Kane, P.M., Holowka, D., Baird, B. 1988. Cross-linking of IgE-receptor complexes by rigid bivalent antigens >200 A in length triggers cellular degranulation. J. Cell Biol107: 969.

Lider, O., Santos, L.M.B., Lee, C.S.Y., Higgins, P.J., Weiner, H.L. 1989. Suppression of experimental autoimmune encephalomyelitis by oral administration of myelin basi protein. 11. Suppression of disease and in vitro immune responses is mediated by antigen-specific CD8+ T lymphocytes. J. Immunol. 142: 748.

Liener, I.E. 1980. Toxic Constituents Of Plant Foodstuffs, 2nd. ed. Academic Press, New York, NY.

Marsh, M.N. 1992. Gluten, major histocompatibility complex and the small intestine: a molecular and immunobiologic approach to the spectrum of gluten sensitivity ('celiac sprue') Gastroenterology 102: 330.

Metcalfe, D.D. 1992. The nature and mechanisms of food allergies and related diseases. Food Technol. 46: 136.

Miller A., Lider, O., Weiner, H.L. 1991. Antigen-driven bystander suppression after oral administration of antigens. J. Exp. Med. 174: 791.

Mosmann, T.R. and Coffman, R. L. 1989. TH1 and TH2 cells: different patterns of lymphokine secretion lead to different functional properties. Ann. Rev. Immunol.

Munro, H.N., and Crim, M.C. 1980. The proteins and amino acids. Ch. 3 In Modern Nutrition in Health and Disease, R.S. Goodhart and M.E. Shils (Ed.), p. 82-83. Lea & Febiger, Philadelphia, PA.

Ngan, J. and Kind, L.S. 1978. Suppressor T cells for IgE and IgG in Peyer's patches of mice made tolerant by the oral administration of ovalbumin. J. Immunol. 120: 861.

O'Farrelly, C. and Gallagher, R.B. 1992. Intestinal gluten sensitivity: snapshots of an unusual auto-immune like disease. Immunol. Today 13: 474.

Perkin, J. E. 1990. Maternal influences on the development of food allergy in the infant. Ch. 5. In Food Allergies and Adverse Reactions, J.E. Perkin (Ed.), p. 89-94, p. 103. Aspen Publishers, Inc., Gaithersburg, MD.

Peters, T.J., Jones, P.E., Wells, G. 1978. Analytical subcellular fractionation of jejunal biopsy specimens: enzyme activities, organelle pathology and response to gluten withdrawal in patients with coeliac disease. Clin. Sci-Mol Med. 55: 285.

Raulet, D.H. 1989. Antigens for gamma/delta T cells. Nature. 339: 342.

Richman, L.K., Graeff, A.S., Yarchoan, R., Strober, W. 1981. Simultaneous induction of antigen-specific IgA helper T cells and IgG suppressor T cells in the murine Peyer's patches after protein feeding. J. Immunol. 126: 2079.

Sachs, J.A., Awad, J., McCloskey, D., Navarrette, C., Festenstein, H., Elliot, E., Walker-Smith, J.A., Griffiths, C.E., Leonard, J.N., Frye,L. 1986. Different HLA associated gene combinations contribute to susceptibility for coeliac disease and dermatitis herpetiformis. Gut 27: 515.

Sampson, H.A. 1992 Food hypersensitivity: manifestations, diagnosis, and natural history. Food Technol. 46: 141.

Sampson, H.A. and Metcalfe, D.D. 1991. Immediate reactions to foods. Ch. 7. In Food Allergy: Adverse Reactions to Foods and Food Additives, D.D. Metcalfe, H.A.

Sampson and R.A. Simon (Ed.), p. 99. Blackwell Scientific Publications, Inc., Boston,MA.

Schild, H., Mavaddat, N., Litzenberger, C., Ehrich, E.W., Davis, M.M., Bluestone, J.A., Matis, L., Draper, R.K., Chien, Y-h. 1994. The nature of the major histocompatibility complex recognition by gamma delta T cells Cell. 76: 29. Sloan, A.E. and Powers, M.E. 1986. A perspective on popular perceptions of adverse reactions to foods J. Allergy Clin. Immunol. 78: 127.

Strobel, S. and Ferguson, A. 1984. Immune responses to fed protein antigens in mice. 3. Systemic tolerance or priming is related to age at which antigen is first encountered. J. Pediatr. Res. 18: 588.

Strominger, J.L. 1989. The gamma delta T cell receptor and class Ib MHC-related proteins: enigmatic molecules of immune recognition. Cell 57: 895

Taylor, S.L. 1992. Chemistry and detection of food allergens. Food Technol. 46: 146. Trejdosiewicz, L.K., Calabrese, A., Smart, C.J., Oakes, D.J., Howdle, P.D., Crabtree, J.E., Losowsky, M.S., Lancaster, F., Boylston, A.W. 1991 Gamma delta T cell receptorpositive cells of the human gastrointestinal mucosa: occurence and V region gene expression in Heliobacter pylori-associated gastritis, coeliac disease and inflammatory bowel disease. Clin, Exp. Immunol. 84: 440.

Troncone, R. and Auricchio, S. 1991. Gluten sensitive enteropathy (celiac disease). Food Rev. Internat. 7: 205.

Vitoria, J.C., Camarero, C., Sojo, A., Ruiz, A., Rodriguez-Soriano, J. 1982. Enteropathy related to fish, rice, and chicken. Arch. Dis. Child. 57:44.

Weiner, H.L., Friedman, A., Miller, A., Khoury, S.J., Al-Sabbagh, A., Santos, L., Sayegh, M., Nussenblatt, R.B., Trentham, D.E., Hafler, D.A. 1994. Oral tolerance: immunologic mechanisms and treatment of animal and human organ-specific autoimmune diseases by oral administration of autoantigens. 1994. Annu. Rev. Immunol. 12: 809.

Weiser, M.M. and Douglas, A.P. 1976. An alternative mechanism for gluten toxicity in coeliac disease. Lancet 1(7959): 567.

MICROCIRCULATION, VITAMIN E AND OMEGA 3 FATTY ACIDS: AN OVERVIEW

Geza Bruckner

Department of Clincal Sciences/Division Clinical Nutrition
University of Kentucky
Lexington, KY 40506-0080

INTRODUCTION

Fatty Acid Isomers and Antioxidants

Epidemiological and clinical studies suggest that ingestion of diets rich in fish and fish oils (FO) may be related to reduced incidence of cardiovascular disease (CVD). Our studies [1,2] and many other studies, in humans and using animal models for atherosclerosis, have shown that FOs significantly reduce serum lipoproteins (in particular very low density lipoproteins-VLDL), and decreases platelet reactivity [3,4,5,6,7,8,9]. It has been recently demonstrated that the hypotriglyceridemic effect may be due to decreased hepatic mRNA encoding for fatty acid synthase [10]. However, not all reported effects of n-3 fatty acids appear to be beneficial [11,12,13,14]. Fatty acids of the three series are susceptible to rapid oxidation and the resulting free radicals can elicit cellular damage [15]. In this regard, a poor antioxidant status has been found to be associated with an increased risk for CVD [16,17,18]. Some studies have reported that FO consumption increases the dietary vit. E requirement [16,19,20,21] while a study recently conducted by Berlin et al. [22] found that dietary FO vs a blend of stripped lard, beef tallow and corn oil supplementation increased red blood cell (RBC) membrane vit. E concentration; surprisingly, plasma and tissue lipid oxidation products were not measured. In line with these observations, Chen et al [23] reported that FO supplementation may in fact attenuate free radical generation (measured as MDA) in rabbits following coronary occlusion, presumably by serving as the substrate for peroxidation at the expense of other membrane constituents. They also found that in the FO fed group the level of myocardial superoxide dismutase and the 6-keto-PGF$_1$ α/TXB$_2$ ratio increased versus a coconut oil control (FO vs chow fed did not differ significantly). An alternative explanation might be related to the decreased TXB$_2$ which results in equimolar reductions in HHT and MDA. Most likely, as suggested by others [21], the oxidative stress induced by high levels of dietary FO can be countered by vit. E supplementation in animals where other antioxidant mechanisms are not compromised, although the amounts required are unclear.

Lipid/Lipoprotein oxidation: Role of Vitamin E

Lipoprotein oxidation results in numerous physical/chemical changes, e.g., increased electrophoretic mobility [24], degradation of associated apoproteins[25], loss of vit. E and polyunsaturated fatty acids (PUFA) [26] and increased levels of lipid oxidation products [27]. It also results in numerous biological effects, e.g., cytotoxicity [27,28] altered production of mediators like platelet derived growth factor (PDGF), interleukin (IL-1), and prostacylin (PGI$_2$) [29,30,31,32,33], and is chemotactic[34]. Lehr et al[35] reported that oxidized low density lipoprotein (oxLDL) stimulates leukocyte adhesion <u>in vivo</u> through a mechanism that depends on both the action of leukotrienes and superoxide radicals and involves platelet-activating factor (PAF) receptors or PAF-like lipids that are generated in a free radical-catalyzed peroxidation of phospholipids. Recently, Faruqi et al.[36] have demonstrated that vit. E addition to human endothelial cell cultures decreased monocyte adhesion most likely via the observed concomitant decrease in E-selectin mRNA.

Consumption of antioxidants is thought to protect against atherogenesis. In fact, addition of antioxidants, including vit. E, or chelators of transition metals prevented <u>in vitro</u> modification of LDL by cells [37,38]. However, under conditions whereby the α tocopherol radical is not eliminated, as in the absence of vitamin C or during rapid lipid oxidation, the tocopherol radical can propagate a PUFA radical chain reaction in the LDL particle [39] and these radical initiated LDL modification reactions may be inhibited <u>in vitro</u> by superoxide dismutase [40]. Furthermore, oxy radical scavengers, e.g. superoxide dismutase, catalase and dimethylsulfoxide have been shown in some cases to attenuate ischemia-reperfusion (I/R) injury [41]. Other antioxidants which may protect against LDL oxidation include ß-carotene, BHT, probucol, lycopene, and cryptoxanthine [42,43,44]. However, there are conflicting reports as to whether ß-carotene serves as an effective <u>in vivo</u> antioxidant [45]. A recent review of the epidemiologic and clinical evidence [46], suggests that supplemental vit. E reduces CVD risk by approximately 30 - 40%.

The generation of additional oxLDL or other lipid oxides could lead to increased endothelial damage and subsequent increased adhesion of platelets and leukocytes, resulting in enhancement of endothelial permeability, increased intimal hyperplasia and accelerated atherogenesis. The n3 fatty acids in the presence of adequate antioxidant are postulated to attenuate this cascade of events.

Cardiovascular Disease and Fatty Acids

The etiology of CVD is multifaceted and although many findings support the involvement of dietary fat and cholesterol in atherogenesis [47,48,49], the mechanisms involved are not clearly understood. For example, the amount of dietary fat is directly correlated to the serum cholesterol concentration, however, different fatty acid isomers, e.g. n3 vs. n6 20 carbon fatty acids, can elicit completely different vascular responses as evidenced by the Greenland Eskimo studies [50] and others [51,52,53,54]. These findings suggest that n3 fatty acids markedly reduce the incidence of CVD but do not lower serum cholestrol levels like the n6 fatty acids. Work by Grundy [55], and others [56,57,58], also suggests that various monounsaturated fatty acids (MONO) have different cholesterol-lowering effects. Serum cholestrol levels can not solely explain the effects of various dietary fatty acid isomers on the etiology of CVD. As recently demonstrated in a large ecological study of 65 counties in China [59], erythrocyte MONO were found to be inversely related to the incidence of CVD. While plasma cholesterol levels were not correlated with CVD, given the likely role of oxLDL in atherogenesis these data suggest that MONO reduce oxidation of LDL or other lipids and thus provide a protective effect [60]. Coronary lesions in different coronary arteries progress at near-independent rates despite being exposed to the same concentrations of serum lipoprotein concentrations [61] thus, suggesting that blood flow/shear stress or other factors may contribute significantly to the disease process.

One prevalent manifestation of CVD is peripheral arterial disease (PAD), which is a major cause of CVD mortality and morbidity. PAD as manifested by intermittent claudication affects about 0.5 to 1.0% of individuals above age 35, with a twofold predominance in men. Mortality in patients with intermittent claudication has been shown to be six times higher than in other people of similar age and gender. This excess in mortality is primarily due to a ninefold increase in CVD deaths. The Framingham study [62,] showed a significant correlation between plasma triglyceride levels and PAD, and more recently an association between PAD and the ratios of lipid oxides to total lipids has been suggested [63,].

Possible Vascular Alterations by n3 Fatty Acids

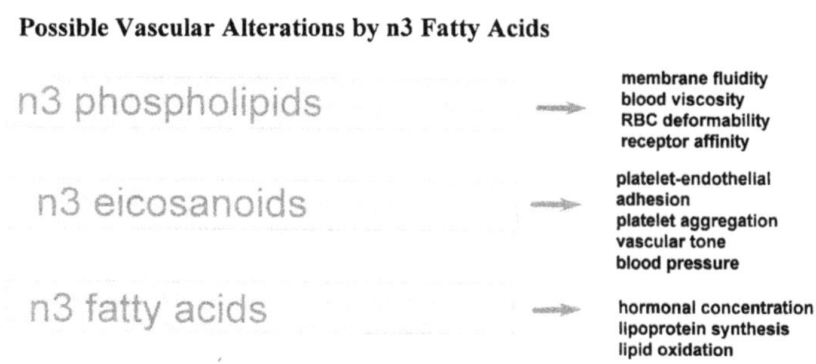

Figure 1. Alterations of biochemical and physiological responses by dietary n3 fatty acids.

Blood Flow/n3 Fatty Acids

Possible mechanisms for vascular alterations induced by n3 fatty acids are depicted in Fig 1. Bruckner et al. have reported that FO supplementation can increase skin blood flow

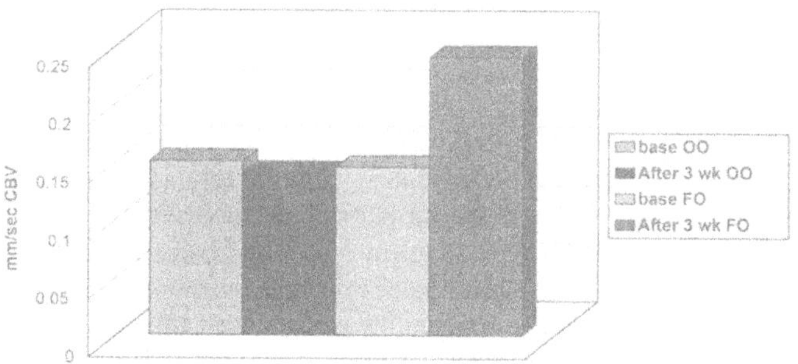

Figure 2. Nailfold capillary blood cell velocities in normolipidemic subjects supplemented with olive oil (OO) or fish oil (FO) for 3 weeks.

in humans (Fig 2) (2). Furthermore, in an elderly population the increased capillary blood cell velocity induced by n3 fatty acids was maintained only in the presence of vitamin E (Fig 3). In a recently completed study with twenty-six hyperlipidemic male subjects aged 18-55, we

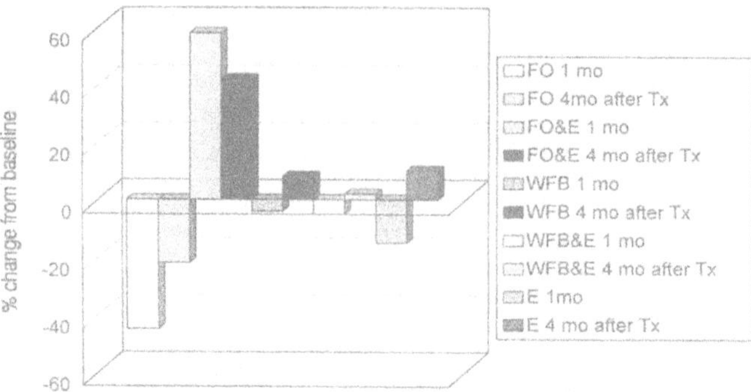

Figure 3. Fats were supplemented for up to 4 mo. FO=fish oil, WFB=western fat blend, E=vitamin E; fat at 1 g/Kg/day, E at 100 IU/10Kg/day.

found that CBV was increased significantly by FO and safflower oil (vit. E, 1 IU/g oil) with both oat bran and wheat bran supplementation over baseline. CBV did not return to baseline values after a one month washout period following either FO and fiber or SO and fiber intervention. There were no significant decreases in wholeblood viscosity (as measured by a capillary viscometer) due to FO. There were no blood pressure changes as a result of n-3 fatty acid supplementation. Therefore it appears that not only the dietary fat but the supplemental fiber sources may interact to influence microcirculatory blood flow. Pscheidl et al. [64,] demonstrated that short term FO administration in a total parenteral nutrition solution vs triglycerides consisting primarily of oleic and linoleic acids significantly improved tissue perfusion after an endotoxin insult (vit. E status was not clearly defined). Lehr et al. [65,] reported that in an ischemic-reperfused striated muscle preparation the RBC velocity was significantly elevated in the FO fed vs control group 24 hrs after the anoxia; furthermore, FO attenuated plasma fibrinogen, leukocyte-endothelial adhesion, capillary obstruction, and capillary leakage (the possible mechanisms of action nor the dietary or tissue antioxidant status was not measured). However, Ming-Fong Chen et al. [66,], while observing reduced atherosclerotic lesions and tissue cholesterol in the aorta and pulmonary artery of FO fed hypercholesterolemic rabbits, did not observe a myocardial protective effect of the FO following acute coronary occlusion-reperfusion (vit. E was added to the oils fed) - this is in contrast to their more recently reported data[23]. Ellis et al.[67,] showed that n3 fatty acids diminished the bradykinin dilator response in cerebral microvessels which is known to be mediated by oxygen radicals (oils were supplemented with minimal amounts of antioxidants). Rat hearts from Wistar-Kyoto and stroke-prone hypertensive rats showed that the animals fed FO vs corn oil had improved cardiac contractility performance after ischemia-reperfusion(I/R) (antioxidant status of the oils fed was not stated) [68,]. Myocardial damage after I/R in rats fed menhaden vs corn oil was significantly decreased using creatine kinase concentration as an index of damage; survival rates were also significantly improved in the menhaden fed group (BHQ was added to the diets but the vit. E levels were not stated)[69,]. Malis et al.[70,], using preanoxic or postanoxic aortic rings from rats fed fish, corn or beef tallow diets, concluded that relaxation to acetylcholine was greater in the FO fed group and others [71,] have shown that after small-artery repair the downstream beneficial effects of dietary FOs is most likely due to diminished vasoconstriction of the vessel. A summary of the n3 fatty acid and blood flow alterations is presented in Table1. Jaye et al. [72,], suggested that the suppressive effect of marine oils on vascular reactivity is in part due to decreased norepinephrine responsiveness of the vessels; it has been shown that n3 fatty acids alter α but not ß-adrenoreceptor activity

[73]. Other recently published studies have shown that FO reduces intimal hyperplasia by inhibiting the production of growth factors for smooth muscle cells[29]. However, the addition of vit. E, BHT or ceruloplasmin [74] apparently negates this effect, suggesting that a lipid oxide is needed for the inhibition of platelet-derived growth factor (PDGF).

Although several of the studies related to platelet-leukocyte-endothelial cell interactions appear to show beneficial effects, we postulate that the levels of lipid oxides and inadequate dietary antioxidants in the FO fed animals may in part be responsible for the noted discrepancies regarding physiological and biochemical responses. Our preliminary studies, and others [18,75], suggest that the ratio of dietary antioxidant (vit. E) to the level of n3 fatty acids fed appears to influence the physiological response.

Microcirculation in Disease and Ischemia Reperfusion (I/R)

Changes in capillary blood flow have been noted in subjects with atherosclerotic disease [76], however, it is not clear whether these changes contribute to the etiology of the disease or are the consequence of the disease process. Richardson and Schwartz [77] and others [78] have noted differences in reactive hyperemia responses in the nail fold capillaries of fingers and toes in control subjects versus subjects with peripheral vascular disease; the reactive hyperemia response (time to reach peak capillary velocity) was delayed in diabetic subjects with vascular disease compared to controls [77,79]. Recently, McVeigh et al. [80] have shown that patients with Type-2 diabetes mellitus given FO vs olive oil supplementation significantly improved forearm blood flow responses to acetylcholine, suggesting that the FO stimulates the release of nitric oxide from endothelial cells. The stimulatory effect of n3 fatty acids on NO production in other cells, eg. macrophages, has been found by others [81,82,83]. There is considerable evidence that hypercholesterolemia impairs endothelium dependant relaxation (EDR) and that L-arginine, the NO precursor, restores the EDR response [84]. Changes in blood flow and platelet-leukocyte-endothelial cell interactions, which may be modulated by dietary lipids and antioxidants, have been implicated in directly influencing atherogenesis [85]. As demonstrated by Gibson et al.[86], low shear stress was significantly correlated with an

Table 1	**Summary of Blood Flow Studies and the Effects of Fish Oils**		

Study	Parameter Measured	Outcome	Note
Bruckner et al. (1)	CBV	Increased CBV	n3 vs n9
Ware et al. (2)	LDF	Increased LDF	Vitamin E added
Pscheidl et al. (64)	Tissue perfusion	Improved	Following endotoxin insult
Lehr et al. (65)	RBC velocity	Elevated	Following I/R
McVeigh et al. (80)	Forearm blood flow	Improved	Acetylcholine response (NO?)
Ellis et al. (67)	Microvessel dilation	Diminished	Bradykinin response
Jaye et al. (72)	Forearm vasoconstriction	Inhibits	Acetylcholine response

increased rate of atherosclerosis progression and shear stress appears to be inversely related to endothelin-1 release [87].

It is postulated that the interactions of circulating blood elements with endothelial surfaces depend in part on blood flow/shear stress. However, the constant exposure of the endothelium to various blood components, including prooxidants such as fatty acid oxides or oxidized derivatives of cholesterol, may damage or alter a number of endothelial cell functions, e.g. synthesis of antiaggregatory vasodilators (prostacylin, nitric oxide), endothelial barrier function and/or platelet-leukocyte endothelial cell adhesion. It is possible that these oxidized lipids increase chances for damage to the endothelium in slower flowing vessels. It is important to better understand the effects of different fatty acids not only on lipoprotein metabolism but also how they alter blood flow. Moreover, due to the differing susceptibility of various lipoproteins and fatty acids to oxidative reactions, it is important to determine which of these circulating and membrane lipids are oxidized and the effects of these events on microcirculatory changes, capillary endothelial integrity and atherogenesis.

Several of the mechanisms currently postulated to be involved with reperfusion injury are similar to mechanisms cited for involvement in occlusive vascular disease, e.g., cytotoxic oxygen derived free radicals, lipid oxides, platelet and leukocyte endothelial adhesion, altered capillary permeability and a compromised cellular antioxidant status.

Blood Cell Interactions With The Endothelium

The ying-yang relationship between leukocytes-platelets and the endothelium are delicately balanced and injury to the endothelium leads to increased platelet-leukcocyte adhesion and ultimately contributes to atherogenesis. The endothelium produces endothelium-derived relaxing factors (EDRF-nitric oxide, prostacyclin) as well as endothelin-1, a potent vasoconstrictor/pressor peptide [88]. Platelets produce the potent aggregatory and vasoconstrictive compounds TXA_2 and platelet derived growth factor (PDGF). - a PDGF-like protein designated PDGFc has also been isolated from endothelial cells. PDGF and PDGFc are immunologically indistinguishable and both induce smooth muscle cell (SMC) proliferation and are SMC chemoattractant. FOs appear to inhibit the production of PDGFc and this effect is attenuated in the presence of antioxidants [29], suggesting that an n3 lipid oxide is the agonist. Vanhoutte et al. [89] reported that n3 fatty acids augment the release of EDRF in isolated blood vessels and that the EDRF released differs from nitric oxide and prostacyclin. Others, as previously mentioned, have demonstrated increased NO production from various tissue elements by FOs [81,82,83]. Histamine, thrombin, bradykinin, leukotriene C_4, and free radicals stimulate the rapid migration of P-selectin to the platelet-endothelial cell surface [90]. Cytokines such as IL-1, TNF and LPS activate the endothelial cell expression of E-selectin, ICAM-1 and VCAM-1. It has been found by some investigators that n3 fatty acids decrease the concentration of cytokines [91,92,93] whereas others have demonstrated an elevated cytokine response [82,94].

Platelet aggregation was reported to be diminished [95] or unchanged [96,97] in humans fed FOs. Owens and Cave [98] and others [99,100] reported that changes in platelet fatty acid composition resulting from FO feeding did not influence their adherence to subendothelial basement membrane. However, 8 weeks of cod liver oil feeding to human males and females reduced thromboplastin activity on peripheral monocytes [101]. Recently, Kim et al [102] demonstrated that FO supplemented swine had significantly reduced _in vivo_ platelet microthrombus formation and adherent monocytes. FO feeding of hamsters inhibited oxLDL-induced increase in leukocyte rolling and sticking to arteriolar and post-capillary venular endothelium [103]. Furthermore, vit. E has been shown to diminish monocyte and platelet adhesion to a variety of adhesive proteins when tested at low shear rate in a laminar flow chamber [104]. Thus, the effects of dietary n-3 fatty acids on _in vivo_ platelet-leukocyte-endothelial cell interactions are currently unclear and the noted discepancies may be related to the differing amounts of antioxidants present in the diets fed. Therefore, the ratios of dietary fatty acids to vit. E may determine whether there is a positive or negative effect on platelet-leukocyte-endothelial cell adhesion.

Mouse Preliminary Ongoing Studies: A) Lipid/Antioxidant Interactions B) Ear Model Development C) Platelet-Endothelial Interactions

A) Lipid Antioxidant Interactions

Swiss-Webster mice (24 - 6/group) were fed an AIN-76 diet with 10% of the fat as a) Fish oil (F) b) F & vit. E-100 IU/Kg diet(F+E), c) stripped lard (L), d) stripped lard & vit. E (L+E) e) hydrogenated lard (HL) or f) hydrogenated lard & vit. E (HL+E) for 4 weeks; all diets contained 2% linoleic acid. Following the dietary intervention RBC/MDA at 0,1 and 2

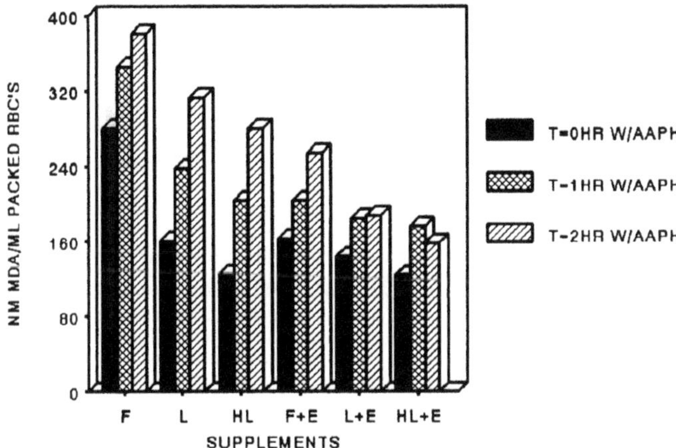

Figure 4. RBC-MDA levels after 0, 1, and 2 hours incubation with AAPH in mice supplemented with oil for 4 weeks, with or without vitamin E.

hours exposure to AAPH were measured (See Figure 4). The data show that F vs L or HL feeding imposes a major oxidative stress as evidenced by highly significant increases of plasma MDA. The addition of 100 IU/Kg vit. E to the F vs L or HL diets does not normalize plasma MDA. AAPH induced hemolysis shows similar trends. The MDA generated by RBC during a two hour incubation indicates that this amount of vit.E cannot normalize the F vs L or HL diets (at 10% fat). Furthermore, the greater the amount of unsaturated fatty acids in the diet the higher the MDA concentrations in both plasma or RBC.

B) Mouse Ear Model Development

To better understand the effects of n3 fatty acids and antioxidants on the mechanism(s) involved in platelet-leukocyte-endothelial cell interactions, we have developed the mouse ear microcirculatory model to investigate these events. The ear vasculature is easily visualized using videomicroscopy and enables us to perform the following measurements: CBV and blood flow in 30-40 µm diameter arterioles and venules, platelet and leukocyte adherence to endothelium, thrombus formation, vessel diameter changes, vessel occlusion and endothelial permeability (Figure 5).

Forty-eight male, 8 week old, Swiss-Webster mice were randomly allocated into three treatment groups (12/group) as follows: a) Mixed Control, equal parts safflower, linseed and olive oils; b) Lard (L); c) Fish, 2% safflower oil and 8% F; 100 IU vit. E -acetate/100 g diet.

Mice were anesthetized with sodium-pentobarbital (90 mg/Kg Bwt) and placed upon a

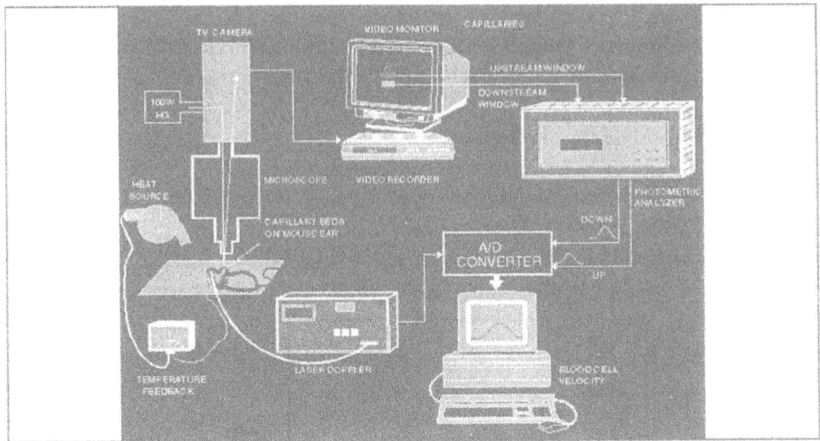

Figure 5. Videomicroscopy methodology to study platelet-endothelial cell interactions in a mouse ear capillary model.

modified microscope stage. The right ear was placed upon a raised lucite block to which was affixed a temperature probe. The mouse ear was taped to the probe and the ear temperature

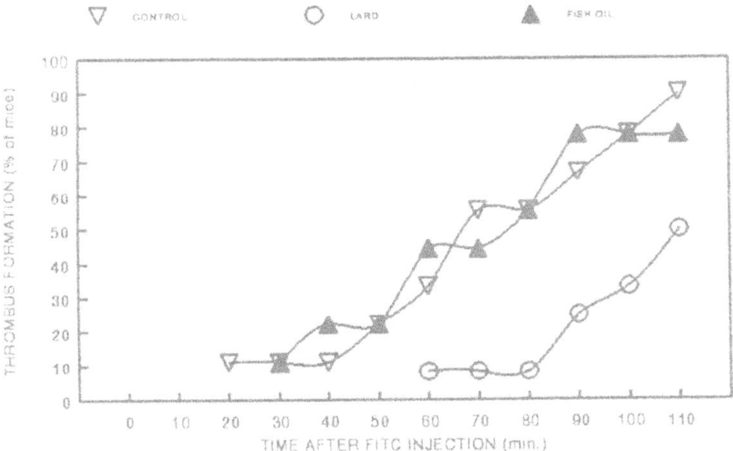

Figure 6. Dietary fat effect on FITC-induced thrombus formation in mice.

was maintained at 32°C with warm air using a feedback control system. Ear temperature was stabilized at 32°C for 5 minutes before measurements were taken. During a 10 minute baseline period, an image of the capillary bed was projected through a microscope (Leitz Laborlux 12 epi-illumination microscope, magnification 200X) to a video monitor and recorded on a VCR for future microcirculation analyses. After baseline recording, fluorescein isothiocyanate-dextrans (FITC MW70,000; Sigma Chemical Co., St.Louis) was injected (50 mg/100 g bwt) into the tail vein. The microscope light filter was changed to provide a wavelength of 490 nm to the capillary field of the mouse ear for 10 min, which caused excitation of FITC thereby initiating free radical mediated damage to the endothelium and other blood elements with subsequent platelet aggregation and thrombus formation. The capillary bed image was recorded on a VCR for a period of two hours or until complete occlusion of the observed vessel. The rate of thrombus formation for the various dietary groups was determined. The antioxidant status of plasma or blood elements was not measured. As shown in Figure 6, the % of mice with thrombi formation, at all time periods after FITC injection, was greater in the more unsaturated fat fed groups. Based on the results previously described , it is plausible that the level of antioxidant (100 IU/Kg) was inadequate to completely prevent oxidation of the more unsaturated fats.

C. Mouse Ear Capillary Integrity (Leakage)

In the above described model, following the free radical mediated injury the FITC leaked from the intra-vascular space into the interstitial tissue over time. In a subset of these animals, we developed a methodology for quantifying the leakage of the FITC over time. The leakage over time was quantified as a change in light intensity using Macintosh Adobe Photoshop Software (Version 2; Apple Computer, Inc.) by capturing the recorded vessel image onto the computer monitor. Vessel images were captured at times which corresponded to microcirculatory analyses (as above). Saved images were then analyzed by use of Screenplay Software (SuperMac Technology Inc.). This program enabled us to quantify changes in the optical density of the recorded image relative to baseline optical densities. We quantified the mean relative light density as the average number of pixels on the video monitor using a density scale of 0 to 255 where 0 is black and 255 is white). The shift from dark to light corresponds to the diffusion of FITC from the vascular lumen to the interstitial space .

Summary

We have observed significant increases in LDF and similar trends for CBV after FO supplementation in younger subjects (both normolipidemic and hyperlipidemic). In elderly subjects, this trend appears to be reversed unless subjects are supplemented with higher doses of vit. E (100 IU/10 KG/day)[1,2]. Our mouse data suggest that dietary vit. E at 100 IU/Kg does not adequately protect against lipid oxidation _in vivo_ or _in vitro_ following an oxidative insult when mice are fed an 8% FO & 2% linoleic acid containing diet.

It has been reported that FO significantly lowers triglycerides and VLDL-cholesterol (especially where subjects have higher initial triglyceride values) and tends to increase LDL-cholesterol and Apo-B100 [3]. These findings are all the more important because the oxidation of LDL from FO-supplemented subjects caused a time-dependent increase in the ability to facilitate albumin transfer which was not diminished following a 2 month washout (WO). Addition of vit. E to the FO supplement prevented this change. These data suggest that FO supplementation without sufficient vit. E may be deleterious to the vascular endothelium. The western fat blend supplement appeared to be protective with increased length of supplementation most likely due to increased MONO fatty acids which are resistant to oxidation[3]; vit. E supplementation appeared to have little additional effect.Our combined studies, and those reported by others, suggest that in humans, increased peripheral microcirculatory flow is most likely due to changes in precapillary vascular tone i.e., vasodilation. It is also possible that subtle changes in each of the three variables i.e. blood pressure, blood viscosity and vascular tone when combined may contribute to the significant changes which we have noted as increased LDF or CBV after intervention with dietary n-3 fatty acids. We hypothesize that interactions between dietary fatty acids and vit. E alters the ratios of vasoconstrictive-platelet aggegatory/vasodilatory-antiplatelet aggregatory agents (TXA_2 and endothelin/PGI_2 and nitric oxide), the expression of adhesion molecules (P-selectin and E-selectin) and thereby directly influences the modulation of free radical mediated events between blood elements and the vascular endothelium. Fatty acids of the n3 series may alter these events by favoring the production of vasodilatory compounds and decreased expression of P and/or E-selectins, provided that these highly oxidizable lipids are protected by adequate antioxidants.

Acknowlegements

A special thanks to all of my collaborating collegues who have helped to advance our knowlege related to the effects of fish oils and cardiovascular events. A special thanks to John Kinsella for his continued support. Supported in part by grants from NIH and Hoffman LaRoche.

REFERENCES:

1.   Bruckner G, Webb P, Greenwell L, Chow C, and Richardson D. Fish oil increases peripheral capillary blood cell velocity in humans. Atherosclerosis. 1987;66:237.

2.   Ware SK, Bruckner G, Atakkaan A, Giles T, Webb P, Chow C, and Richardson DR. Interaction between omega-3 fatty acids, vitamin E, and cutaneous blood flow in healthy elderly male subjects. Nutr Metab Cardiovasc Dis. 1992;2:33.

3.   Harris WS, Connor WE, Inkeles SB, and Illingworth DR. Dietary omega-3 fatty acids prevent carbohydrate-induced hypertriglyceridemia. Metabolism. 1984;33:1016.

4.   Harris HS. Fish oils and plasma lipid and lipoprotein metabolism in humans: a critical review. J Lipid Res. 1989;30:785.

5.   Dyerberg J, and Bang HO. Haemostatic function and platelet polyunsaturated fatty acids in Eskimos. Lancet. 1979;ii:433.

6.   Thorngren M, and Gustafson A. Effects of 11 week increase in dietary eicosapentanoic acid on bleeding time, lipids and platelet aggregation. Lancet. 1981;ii:1190.

7.   Seiss WR, Roth P, Scherer B, Kurzmann I, Bolig B, and Weber PC. Platelet-membrane fatty acids, platelet aggregation, and thromboxane formation during a mackerel diet. Lancet. 1980;i:441.

8.   Kobayashi S, Hirai A, Terano T, Hamazaki T, Tamura Y, and Kumaga A. Reduction in blood viscosity by eicosapentanoic acid. Lancet. 1981;ii:197.

9.   Mortensen JZ, Schmidt EB, Nielson AH, and Dyerberg J. The effect of n-6 and n-3 polyunsaturated fatty acids on hemostasis, blood lipids and blood pressure. Thromb Haemost. 1983;50:543.

10. Jump DB, Clarke SD, Thelen A and Liimatta M. Coordinate regulation of glycolytic and lipogenic gene expression by polyunsaturated fatty acids. J Lipid Res 1994;35:1076.

11. Singer P, Jaeger W, Wirth M, Voigt S, Naumann E, Zimontkowski S, Hajdu I, and Goedicke W. Lipid and blood pressure lowering effect of mackerel diet in man. Atherosclerosis. 1983;49:99.

12. Logani MK, and Davies RE. Lipid oxidation: biological effects and antioxidants - a review. Lipids. 1980;15:485.

13. Flaten H, Hostmark AT, Kierulf P, Lystad E, Trygg K, Bjerkedal T, and Osland A. Fish-oil concentrate: effects on variables related to cardiovascular disease. Am J Clin Nutr. 1990;52:300.

14. Gey KF, Brubacher GB, and Stahelin HB. Plasma levels of antioxidant vitamins in relation to ischemic heart disease and cancer. Am J Clin Nutr. 1987;45:1368.

15. Gudbjarnason S, and Hallgrimsson. The role of myocardial membrane lipids in the development of cardiac necrosis. Acta Med Scand Suppl. 1976; 587:17.

16. Yagi K. Assay for serum lipid peroxide level and its clinical significance. In: Lipid Peroxides in Biology and Medicine. 1984 (K. Yagi, ed. Academic Press: NY) pp 223-224.

17. Rimm EB, Stampfer MJ, Ascherio A, Giovannucci E, Colditz GA, and Willet WC. Vitamin E consumption and the risk of coronary heart disease in men. N. Engl J Med. 1993;328:1444

18. Stampfer MJ, Hennekens CH, Manson JE, Colditz GA, Rosner B, and Willet WC. VitaminE consumption and the risk of coronary disease in women. N Engl J Med. 1993;328:1444.

19. Tatum VC, Changchit C, and Chow CK. Measurement of malonaldehyde by high pressure liquid chromatography with fluorescence detection. Lipids. 1990;25:226.

20. Wolmarans P, Labadarios D, Spinnler Benade AJ, Kotze TJvW, and Louw MEJ. The influence of consuming fatty fish instead of red meat on plasma levels of vitamins A, C and E. Eur J Clin Nutr 1993;47:97.

21. Garrido A, Garate M, Campos R, Villa A, Nieto S, and Valenzuela A. Increased susceptibility of cellular membranes to the induction of oxidative stress after ingestion of high doses of fish oil: effect of aging and protective action of dl-α tocopherol supplementation. J Nutr Biochem. 1993;4:118.

22. Berlin E, Bhathena SJ, Judd JT, Nair PP, Peters RC, Bhagavan HN, Ballard-Barbash R, and Taylor PR. Effects of omega-3 fatty acid and vitamin E supplementation on erythrocyte membrane fluidity, tocopherols, insulin binding, and lipid composition in adult men. J Nutr Biochem 1992;3:392.

23. Chen MF, Hsu HC, Chen WJ, Lee CM, Wu CC, Liau CS and Lee YT. Fish oil supplementation attenuatesfree radical generation in short-term coronary occlusion-reperfusion in cholestrol-fed rabbits. Prostaglandins 1994;47:307.

24. Henriksen T, Mahoney EM, and Steinberg D. Enhanced macrophage degradation of biologically modified low density lipoprotein. Arteriosclerosis. 1983;3:149.

25. Fong LG, Parthasarathy S, Witztum JL, and Steinberg D. Nonenzymatic oxidative cleavage of peptide bonds in apoprotein B-100. J Lipid Res. 1987;28:1466.

26. Esterbauer H, Jurgens G, Quehenberger O, and Koller E. Autoxidation of human low density lipoprotein: loss of polyunsaturated fatty acids and vitamin E and generation of aldehydes. J Lipid Res. 1987;28:495.

27. Morel DW, Hessler JR, and Chisolm GM. Low density lipoprotein cytotoxicity induced by free radical peroxidation of lipid. J Lipid Res. 1983;24:1070.

28. Hessler JR, Morel DW, Lewis LJ, and Chisolm GM. Lipoprotein oxidation and lipoprotein-induced cytotoxicity. Arteriosclerosis. 1983;3:215.

29. Fox PL, and DiCorleto PE. Fish oils inhibit endothelial cell production of platelet-derived growth factor-like protein. Science. 1988;241:453.

30. Steinberg, D. Antioxidants in the prevention of atherosclerosis. Symposium on atherosclerosis (sponsored by The National Heart, Lung and Blood Institute). Bethesda, Maryland, October 15, 1992. In: Circulation 1992;85:2337.

31. Triau JE, Meydani SN, and Schaefer EJ. Oxidized low density lipoprotein stimulates prostacyclin production by adult human vascular endothelial cells. Arteriosclerosis. 1988;8:810.

32. Rajavashisth TB, Andalibi A, Territo MC, Berliner JA, Navab M, Fogelman AM, and Lusis AJ. Induction of endothelial cell expression of granulocyte and macrophage colony-stimulating factors by modified low-density lipoproteins. Nature. 1990;344:254.

33. Peng S, Hu B, Peng AY, and Morin RJ. Effect of cholesterol oxides on prostacyclin production and platlet adhesion. Artery 1993;20:122-134.

34. Berliner JA, Territo MC, Sevanian A, Ramin S, Kim JA, Bamshad B, Esterson M, and Fogelman AM. Minimally modified low density lipoprotein stimulates monocyte endothelial interactions. J Clin Invest. 1990;85:1260.

35. Lehr HA, Seemuller J. Hubner C, Menger MD, and Messmer K. Oxidized LDL-induced leukocyte/endothelium interaction in vivo involves the receptor for platelet-activating factor. Arteriolsclerosis and Thrombosis 1993;13;1013.

36. Faruqi R, de la Motte C and DiCorleto PE. Alpha-tocopherol inhibits agonist-induced monocytic cell adhesion to cultured human endothelial cells. J Clin Invest. 1994;94:592.

37. Henriksen T, Mahoney EM, and Steinberg D. Enhanced macrophage degradation of low density lipoprotein previously incubated with cultured endothelial cells: Recognition by receptors for acetylated low density lipoproteins. Proc Natl Acad Sci USA. 1981;78:6499.

38. Heinecke JW, Baker L, Rosen H, and Chait A. Superoxide-mediated modification of low density lipoprotein by arterial smooth muscle cells. J Clin Invest. 1986;77:757.

39. Bowry VW, Stocker R. Tocopherol-mediated peroxidation. The prooxidant effect of vitamin E on the radical-initiated oxidation of human low-density lipoprotein. J Am Chem Soc 1993;115:6029-6044.

40. Steinbrecher, UP. Role of superoxide in endothelial-cell modification of low-density lipoproteins. Biochim Biophys Acta. 1988;959:20.

41. Korthuis RJ, Granger DN, Townsley MI, and Taylor AE. The role of oxygen-derived free radicals in ischemia-induced increases in canine skeletal muscle vascular permeability. Circ Res. 1985;57:599.

42. Esterbauer H, Dieber-Rotheneder M, Waeg G, Puhl H, and Tatzber F. Endogenous antioxidants and lipoprotein oxidation. Biochem Soc Trans. 1990;18:1059.

43. Knipping G, Rotheneder M, Striegl G, and Esterbauer H. Antioxidants and resistance against oxidation of porcine LDL subfractions. J Lipid Res. 1990;31:1965.

44. Esterbauer H, Dieber-Rotheneder M, Striegl G, and Waeg G. Role of vitamin E in preventing the oxidation of low-density lipoprotein. Am J Clin Nutr. 1991;53:314S.

45. Reaven PD, Khouw A, Belz WF, Parthasarathy S, and Wiztum JL. Effect of antioxidant conbinations in humans. Arteriosclerosis and Thrombosis 1993;13:590-600.

46. Hennekens CH, Gaziano JM. Antioxidants and heart disease: Epidemiology and clinical evidence. Clin

Cardiol 1993;16:110.

47. Grundy SM, Bilheimer D, Blackburn H, Brown WJ, Kwiterovich PO, Mattson F, Shonfeld G, and Weidman WN. Rationale of the diet-heart statement of the American Heart Association. Report of Nutrition Committee. Circulation. 1982;65:839.

48. Lipid Research Clinics. In: Coronary primary prevention trial results. JAMA 251:351.

49. Zollner N, and Tato F. Fatty acid composition of the diet: impact on serum lipids and atherosclerosis. Clinical Investigator. 1992;70:968.

50. Bang HO, Dyerberg J, and Brondum Nielsen A. Plasma lipid and lipoprotein pattern in Greenlandic West-coast Eskimos. Lancet. 1971;ii:1143.

51. Sanders TAB, Vickers M, and Haines AP. Effect on blood lipids and haemostasis of a supplement of cod-liver oil rich in eicosapentaenoic acid and docosahexanoic acid, in healthy young men. Clin Sci 1981;61:317.

52. Fehily AM, Burr ML, Phillips KM, and Deadman NM. The effect of fatty fish on plasma lipid and lipoprotein concentrations. Am J Clin Nutr. 1983;38:349.

53. Von Lossonczy TO, Ruiter A, Hermus RJJ, Van Gent CM, and Bronsqeest-Schoute HC. Effect of a fish diet on serum lipids in healthy human subjects. Am J Clin Nutr. 1978;31:1340.

54. Illingworth DR, Harris WS, and Connor WE. Inhibition of low density lipoprotein synthesis by dietary omega-3 fatty acids in humans. Arteriosclerosis. 1984;4:270.

55. Grundy, S.M. Comparison of monounsaturated fatty acids and carbohydrates for lowering plasma cholesterol. NEJM. 1986;314:745.

56. Castelli, WP. The triglyceride issue: A view from Framingham. Am Heart J. 1986;112:432.

57. Zock PL and Katan MB> Hydrogenation alternatives: Effects of trans fatty acids and stearic acid versus linoleic acid on serum lipids and lipoproteins in humans. J. Lipid Res 1992;33:399.

58. Lichtenstein AH, Ausman LM, Carrasco W, Jenner JL, Ordovos JM and Schaefer EJ. Hydrogenation impairs the hypolipidemic effect of corn oil in humans. Arteriosclerosis and Thrombosis 1993;14:154.

59. Chen J, Campbell TC, Li JY, and Peto R. Diet, lifestyle and mortlity in China. Oxford, UK: OxfordUniversity Press, 1990.

60. Parthasarathy S, Khoo JC, Miller E. Low density lipoprotein rich in oleic acid is protected against oxidative modification: implications for dietary prevention of atherosclerosis. Proc Natl Acad Sci USA 1990;87:3894

61. Gibson CM, Sandor T, Stone PH, Pasternak RC and Sacks FM. Quantitative angiographic and statistical methods to assess serial changes in coronary luminal diameter and implications for atherosclerosis regression trials. Am J Cardiol 1992;69:1286.

62. Castelli, WP. The triglyceride issue: A view from Framingham. Am Heart J. 1986;112:432.

63. Stringer MD, and Kakkar VV. Markers of disease severity in peripheral atherosclerosis. European J Vasc Surg. 1990;4(5):513.

64. Pscheidl EM, Wan JM, Blackburn GL, Bistrian BR, and Istfan NW. Influence of n-3 fatty acids on splanchnic blood flow and lactate metabolism in an endotoxemic rat model. Metabolism. 1992;41:698.

65. Lehr H, Hubner C, Nolte D, Kohlschutter A, and Messmer K. Dietary fish oil blocks the microcirculatory manifestations of ischemia-reperfusion injury in striated muscle in hamsters. Proc Natl Acad Sci. 1991;88:6726.

66. Chen M, Lee Y, Hsu H, Yeh P, Liau C, and Huang P. Effects of dietary supplementation with fish oil in prostanoid metabolism during acute coronary occlusion with or without reperfusion in diet-induced hypercholesterolemic rabbits. International Journal of Cardiology. 1992;36:297.

67. Ellis EF, Police RJ, Dodson LY, McKinney JS, and Holt SA. The effects of dietary n-3 fatty acids on the cerebral microcirculation. Am J Physiol. 1992;262:H1379.

68. Paulson DJ, Smith JM, Zhao J, and Bowman J. Effects of dietary fish oil on myocardial ischemic/reperfusion injury of Wistar Kyoto and stroke-prone spontaneously hypertensive rats. Metabolism. 1992;41:533.

69. Hock CE, Beck LD, Bodine RC and Reibel DK. Influence of dietary n-3 fatty acids on myocardial ischemia and reperfusion. Am. J. Physiol. 1990;26:H1518.

70. Malis CD, Leaf A, Varadarajan GS, Newell JB, Weber PC, Force T, and Boventre JV. Effects of dietary omega 3 fatty acids on vascular contractility in preanoxic and postanoxic aortic rings. Circulation. 1991;84:1393.

71. Barker JH, Gu JM, Anderson GL, O'Saughnessy M, Pierangeli S, Johnson P, Galleti G., and Acland, RD. The effects of heparin and dietary fish oil on embolic events and the microcirculation downstream from a small artery repair. Plast Reconst Surg;1993;91(2):335.

72. Jaye PF, Gust AP, Nestel PJ and Dart AM. Marine oils dose-dependently inhibit vasoconstriction in forearm resistance vessels in humans. Hypertension 1993;21:22.

73. Reibel DK, Holahan MA and Hock CE. Effects of dietary fish oil on cardiac responsiveness to adrenoceptor stimulation. Am. J. Physiol. 1988;254:H494.

74. Fox PL. Serum factor that blocks the action of fish oils in endothelial cells production of platelet-derived

growth factor is ceruloplasmin. Lab Invest. 1992;66:467.

75. Driskell WJ, Neese JW, Bryant CC, and Bashor MM. Measurement of vitamin A and vitamin E in human serum by high-performance liquid chromatography. J Chromatography. 1982;231:439.

76. Schwartz RW, Freedman AM, Richardson DR, Hyde GL, Griffen WO, Vincent DG, and Price MA. Capillary blood flow: videodensitometry in the atherosclerotic patient. J Vas Surg. 1984;1:800.

77. Richardson D, and Schwartz R. Comparison of resting capillary flow dynamics in the finger and toenailfolds. Micro Endothelium Lymph. 1984;1:645.

78. Fagrell B, Hermannsson IL, Karlander SG, and Ostergren J. Capillary microscopy for assessment of skin viability and microangiopathy in patients with Diabetes Mellitus. Acta Med Scand Suppl. 1984;687:25.

79. Tooke JE, Lins PE, Ostergren J, Fagrell B. Synchronous assessment of human skin microcirculation by laser Doppler flowmetry and dynamic capillaroscopy. Int J Microcirculation Clin Exp. 1984;4:249.

80. McVeigh GE, Brennan GM, Johnston GD, McDermott BJ, McGrath LT, Henry WR, Andrews JW and Hayes JR. Dietary fish oil augments nitric oxide production or release in patients with Type-2 (non-insulin-dependent) diabetes mellitus. Diabetologia 1993;36:33.

81. Shimokawa H and PM VanHoutte. Dietary omega-3 fatty acids and endothelium dependent relaxations in porcine coronary arteries. Am. J. Physiol. 1989;256:H968.

82. Chaet MS, Garcia VF, Arya G, Ziegler MM. Dietary fish oil enhances macrophage production of nitric oxide. J. Surg. Res. 1994;57:65.

83. Schini VB, Durante W, Catovsky S and Vanhoutte PM. Eicosapentaenoic acid potentiates the production of nitric oxide evoked by interleukin-1ß in cultured vascular smooth muscle cells. J Vasc Res 1993;30:209.

84. Schuschke DA, Miller FN, Lominadze DG and Feldhoff RC. L-arginine restores cholestrol-attenuated microvascular responses in the rat cremaster. Int J Microcirc 1994;14:204.

85. Ross R, and Glomset J. The pathogenesis of atherosclerosis. NEJM. 1976;295:369.

86. Gibson CM, Diaz L, Kandarpa K, Sacks FM, Pasternak RC, Sandor T, Feldman C, and Stone PH. Relation of vessel wall shear stress to atherosclerosis progression in human coronary arteries. Arterioclersosis and Thrombosis 1993;13:310.

87. Kuchan MJ and Frangos JA. Shear stress regulates endothelin-1 release via protein kinase C and cGMP in cultured endothelial cells. Am J Physiol 1993;264:H150.

88. Yanagisawa M, Kurihara H, Kimura S, Tomobe Y, Kobayashi M, Mitsui Y, Yazaki Y, Goto K, and Masaki T. A novel potent vasoconstrictor peptide produced by vascular endothelial cells. Nature. 1988;332:441.

89. Vanhoutte PM, Shimokawa H, Boulanger CH. (1991). In Health Effects of w3 Polyunsaturated Fatty Acids in Seafoods. World Review Nutr Diet. (A.P. Simopoulos, R.R. Kiferr, R.E. Martin, and S.M. Barlow, eds., Karger:Basel) vol. 66, pp 233-244.

90. Albelda SM, Smith CW and Ward PA. Adhesion molecules and inflammatory injury. FASEB J. 1994;8:504.

91. Chandrasekar B and Fernandez G. Decreased pro-inflamatory cytokines and increased antioxidant enzyme gene expression by omega-3 lipids in murine lupus nephritis. Biochem Biophys Res Commun. 1994;200(2):893.

92. Endres S, Ghorbani R, Kelley VE, Georgilis K. The effect of dietary supplementation with n-3 polyunsaturated fatty acids on the synthesis of interleukin-1 and tumor necrosis factor by mononuclear cells. N Engl J Med. 1989;320:265.

93. Meydani SN, Lichtenstein AH, Cornwall S, Meydani M, Goldin BR, Rasmussen H, Dinarello CA and Schaefer EJ. Immunologic effects of national cholesterol education panel step-2 diets with and without fish-derived N-3 fatty acid enrichment. J Clin Invest. 1993;92:105.

94. Somers SD and Erickson KL. Alteration of tumor necrosis factor-α production by macrophages from mice fed diets high in eicosapentaenoic and docosahexaenoic fatty acids. Cell Immun 1994;153:287.

95. Skeaff CM, and Holub BJ. The effect of fish oil consumption on platelet aggregation responses in washed human platelet suspensions. Thrombosis Res. 1988;51:105.

96. Rao GHR, Kishore NP, Peller JD, and White JG. (1987). Influence of polyenoic acids on arachidonic acid metabolism and platelet function. In Cardiovascular Disease (L.L. Gallo, ed., Plenum Press:New York) pp 495

97. Rao GHR, Radha E, and White JG. Effect of docosahexaenoic acid (DHA) on arachidonic acid metabolism and platelet function. Biochem Biophys Res Commun. 1985;131:50.

98. Owens MR, and Cave WT. Dietary fish lipids do not diminish platelet adhesion to subendothelium. Brit J Haematol. 1990;75:82.

99. Heizer ML, McKinney JS and Ellis EF. The effect of dietary n-3 fatty acids on in vivo platelet aggregation in the cerebral microcirculation. Throm Res 1992;68:383.

100. Westerveld HT, de Graaf JC, van Breugel HH, Akkerman JW, Sixma JJ, Erkelens DW and Banga JD. Effects of low-dose EPA-E on glycemic control, lipid profile, lipoprotein(a), platelet aggregation, viscosity, and platelet vessel wall interaction in NIDDM. Diabetes Care 1993;16:683.

101. Hansen JB, Olsen JO, Wilsgard L, and Osterud B. Effects of dietary supplementation with cod liver oil on

monocyte thromboplastin synthesis, coagulation, and fibrinolysis. J Int Med. 225 suppl. 1989;1:133.

102. Kim DN, Schmee J, Baker JE, Lunden GM, Sheehan CE, Lee CS, Eastman A, Solis O, Ross JS and Thomas WA. Dietary fish oil reduces microthrombi over atherosclerotic lesions in hyperlipidemic swine in the absence of plasma cholesterol reduction. Exp Mol Pathol. 1993;59(2):122.

103. Lehr HA, Hubner C, Finckh B, Nolte D, Beisiegel U, Kohlschutter A, and Messmer K. Dietary fish oil reduces leukocyte/endothelium interaction following systemic administration of oxidatively modified low density lipoprotein. Circulation. 1991;84:1725.

104. Steiner M. Vitamin E: More than an antioxidant. Clin Cardiol 1993;16:I.

INDEX

GPSR Compliance
The European Union's (EU) General Product Safety Regulation (GPSR) is a set
of rules that requires consumer products to be safe and our obligations to
ensure this.

If you have any concerns about our products, you can contact us on

ProductSafety@springernature.com

In case Publisher is established outside the EU, the EU authorized
representative is:

Springer Nature Customer Service Center GmbH
Europaplatz 3
69115 Heidelberg, Germany

www.ingramcontent.com/pod-product-compliance
Ingram Content Group UK Ltd.
Pitfield, Milton Keynes, MK11 3LW, UK
UKHW020917130726
13719UKWH00012B/167